Mathematical Encounters

Mathematical Encounters

For the inquisitive mind

Paul Chika Emekwulu

ISBN:	Softcover	978-1-4535-5102-8
	Ebook	978-1-4535-5103-5

This book was printed in the United States of America.

To order additional copies of this book, contact:
Xlibris Corporation
1-888-795-4274
www.Xlibris.com
Orders@Xlibris.com
79895

CONTENTS

Author's Preface

To some people, everything about this book may be unusual, weird or strange. You can say that the content is unusual because it is not written strictly in line with any traditional curriculum. You can also say that the chapter lengths are strange. Why? Because some are unusually long, and some are unusually short. You can also say that the general format of the book is unusual and you are right. It is planned that way. Each chapter has a set of objectives and a chapter introduction. Because of the 'strangeness' of this book, I have decided to make the preface a long one. The preface details my intention. You can also say that the title of this book sounds more esoteric (exclusive) than ordinary. I have just one advice. Get into what the book has in stock for you and forget the title.

At times, some of us display the attitude that any school text not written in line with any traditional or conventional curriculum should not only be banned and dumped but thrown into the Atlantic. Most of these materials fall into the category of supplemental materials or texts. The right supplemental materials in mathematics are analogous to novels and other reading materials in English or just any language be it Korean, Japanese, French, Swahili, Finesse, German, Italian, Spanish, or Portuguese, etc.

Novels, the language of expression notwithstanding, build language skills in the areas of vocabulary, reading and comprehension, spelling, grammar, etc. Similarly, the right supplemental materials in mathematics build vocabulary, computational, language, logical and reasoning skills. They also expose students to the appreciation of the aesthetic side of mathematics. *Mathematical Encounters for the Inquisitive Mind* is not different.

Every writer knowingly or unknowingly through his or her writings, reveals a part of himself or herself to the public. *Mathematical Encounters for the Inquisitive Mind* is a collection of different articles I wrote over the years at different times under different real circumstances and each has some dose of mathematical insights for the inquisitive mind.

While on a famous American TV show one wintry morning of 2006, a renounced American blind musician, Steve Wonder was asked by the host why he has not produced any music recently. In his usual manner of twisting his head from left to right, the 55 year old replied, "You have to experience things in order to create music." Steve is right. Creative writing, fiction or non-fiction, in many ways at times, is similar to creating music. Like Stevie Wonder, I feel the same way for most of the individual articles that make up this book. For the various circumstances that led to the creation of these articles, I not only experienced most of them, I lived them as well.

Mathematical Encounters for the Inquisitive Mind has the following chapters:

Chapter 1 is about an indoor game called 'Line game' and how to play it. Line game was exposed to me when I was a student at Alvan Ikoku College of Education, Owerri, Imo state, Nigeria by a casual group of students who were playing the game purely for entertainment. After the encounter, I started to look at the game beyond just entertainment. I later discovered that the Line game has a potential as a teaching tool. The Line game could be used to teach quite a lot of mathematical concepts such as ratio, percentages, proportion, fractions, decimals etc.

Chapter 2 is about the subject of intuition. (This again sounds weird or strange for a math textbook.) It is wrong to think that any knowledge not gained through a conventional classroom should not only be questioned but rejected outright and thrown out through the window or into the Atlantic. Ultimate reality is beyond the five mundane senses of sight, hearing, taste, touch and smell. The famous American psychologist, Professor William James says that human beings use only ten percent of their mental faculties. This of course, raises a question and the question is: Where is the rest? The rest is hidden. The rest is latent. The rest is dormant. The rest is of course, limited by the five mundane or physical senses.

Powerful thoughts and ideas that have contributed to human progress and civilization (both past and present) have always come to their owners not necessarily while in conscious state of existence but in altered state of consciousness. These include dreams, out of body experiences, near-death experience etc. Such thoughts and ideas in more or less forms, need be captured and written down, otherwise, they could disappear. Even if they are not entirely lost, they might lose their original flavor. Therefore, in order to maintain the present level of civilization, and still further human progress, we have to maintain open minds to these experiences. We all have an inner voice. We just have to improve our listening skills.

The chapter gradually leaves intuition and starts defining Fibonacci numbers as numbers of the form:

1, 1, 2, 3, 5, 8, 13, 21, 34, 55, 89, 144, 233, 377, 610, 987, 1597 ...

and Lucas numbers as numbers of the form:

1, 3, 4, 7, 11, 18, 29, 47, 76, 123, 199, 322, 521, 843, 1364, 2207, 3571 ...

It also looks at some properties of Fibonacci numbers which are either verified by proving them or just validated with examples. Proof by contradiction has been used in one of the proofs. In proof by contradiction, we assume that a false mathematical statement is true and then end up proving it to be false. A search for counter examples is encouraged at the end of the chapter.

Chapter 3 introduces you to my dream journal. Each dream starts with a title and the date it occurred. My dreaming life among others, includes flying, seeing unfamiliar phone numbers, and dreaming of elementary and advanced mathematics (at times in form of lectures, and other times in form of just plain mathematical problems). Not less than 70 such dreams are accounted for, more are still being added as this book is being written and more are expected before publication. Because of this continuity factor, this chapter is expected to be an unusually long one.

Chapter 4 introduces an old Chinese manipulative puzzle. The tangram puzzle was introduced to me while a student at Alvan Ikoku College of Education, Owerri by an American math professor, Mr. Edward Himes. This chapter teaches us how to construct the tangram, it also describes the seven pieces. It ends with a lesson on similarity and a problem for illustration.

Chapter 5 titled, "From Murder Scene to Building and Transforming Word Problems into Simple Equations," is a conversation with a friend. Few minutes into this conversation, we were joined by someone, an American-born who thinks that my Nigerian friend and I have an accent, not only that we have an accent but a thick one. This chapter however, is actually about use of variables in representing unknowns algebraically. It ends with eighteen word problems.

Chapter 6 deals on number patterns. Using examples, it discusses the summing of first n positive odd integers. Proof by mathematical induction is also used in the discussion. The reader, as usual, is advised to search for counter examples.

Chapter 7 introduces elementary problems based on the numbers of the Fibonacci sequence. Six such problems are treated. Proof of one of the Fibonacci properties is also discussed. Using examples, this property is verified.

Chapter 8 takes us more to numbers of the Fibonacci sequence. It talks about the beautiful property of 89, the eleventh term of the sequence.

This chapter also proves the following Fibonacci identities for any four consecutive Fibonacci numbers,

$$u_n, u_{n+1}, u_{n+2}, u_{n+3}:$$

$$\frac{u_{n+3}}{2} - \frac{u_n}{2} = u_{n+1}$$

$$\frac{u_{n+3}}{2} + \frac{u_n}{2} = u_{n+2}$$

Each of the above identities is validated with four examples. A search for counter examples is also encouraged at the end of the chapter.

Actually, Chapter 8 is a rejoinder to an article written by Monte J. Zerger, a professor at Adams State College, Alamosa, Colorado, in the United States of America. The article appeared on page 6 of January 1996 issue of *The Mathematics Teacher*, a professional magazine of The National Council of Teachers of Mathematics.

Chapter 9 explores more number patterns. This chapter includes the following: exploring patterns with triangular numbers, deriving a formula for the nth triangular number, deriving a formula for the nth triangular number using partial sums of first n natural numbers, expressing triangular numbers as partial sums, exploring patterns with Lucas numbers, and exploring patterns with Fibonacci numbers.

Chapter 10 tells about my actual personal experience while being remanded in custody at the Oklahoma City County Jail from April 10 through May 5, 1987 by the US Immigration & Naturalization Service-INS. I entered the United States with a visitor's visa. On December 15, 1984, my visa expired and henceforth, I became an illegal alien and got into trouble with The US Immigration. My life has never been the same since this incident.

Chapter 11 was born while I was in the employ of an outbound telemarketing company that sells DirectTV services to customers throughout the contiguous United States. My employment with this company though a brief one, gave me an opportunity to walk away with a treasure that eventually became a part of something that is very dear to me.

Chapter 12 was born out of an e-mail message posted by an American middle or high school teacher to a newsgroup on the internet. She was asking to know how she can link algebra with geometry. In her own words: *"I am wanting to know if there is any way that I can teach something in algebra and then verify or show how it works with geometry. Thank you for your time and contribution"* (italics mine).

This generated some interest in me and I decided to attempt a contribution. This chapter is as a result of that attempt. The chapter is ended with two practice exercises based on Hero's formula. Hero's formula is used in finding the area of a triangle in terms of the sides.

Chapter 13 was conceived one night while on the job. A thought flashed through my mind and I decided to explore it. Before I realized what was happening, I found myself completing a 130 page book just on triangular numbers.

Chapter 14 introduces a summation problem of a subset of numbers of the Fibonacci sequence and a solution proposed by an imaginary student. The student proposed $u_{2n+1} - 1 + u_{2n}$ as sum of first n Fibonacci numbers. In this chapter, we will see why he is wrong. Actually, the sum of first n Fibonacci numbers is given by $u_{n+2} - 1$ regardless whether n is odd or even. This comes with verification. The chapter ends the discussion with a conclusion.

Chapter 15 contains 18 independent trial questions all based on Fibonacci numbers using *a, b*, and *c* to represent three consecutive terms. The book ends with an appendix that lists seminars developed and presented by the author himself and the extent of the author's internet presence.

Intended Audience for This Book

It is really hard to define an audience for a book of this nature more so when one considers the fact that it is not written in line with any traditional curriculum or syllabus as already mentioned. Though, it does not strictly strive to completely meet for example, West African Examinations Council (WAEC) standards, or the standard document of the National Council of Teachers of Mathematics (NCTM) (a professional US based organization whose membership is open to elementary and middle school teachers, junior high, high school, two year and four-year university math teachers, it still has some relevance. If we are really serious about making mathematics fun and interesting, if we are really serious about giving mathematics its due place, if we are really serious about popularizing mathematics, if we are serious about teaching mathematics more creatively and aggressively too, then this book and similar ones should attract the attention of the following:

National Council of Teachers of Mathematics (NCTM)
Federal Ministry of Education, Abuja, Nigeria
Mathematical Association of Nigeria (MAN)
National Mathematics Center (NMC), Abuja, Nigeria
West African Examinations Council (WAEC)
Nigerian Educational Research & Development Council (NERDC)
Nigerian Science Teachers Association (NSTA)
Nigerian Union of Teachers (NUT)

Nigerian National Petroleum Corporation (NNPC), one of the corporate partners of NMC; AP (Advanced Placement) math students - for college bound high school seniors in US high schools), gifted students participating in an independent study during high school, mathematicians of all levels, those in the business of personal development, researchers, and number theorists, Nigerian state Ministries of Education, principals, math teachers and proprietors of all private secondary schools in Nigeria, all Federal unity secondary schools, math departments of all Nigerian universities, polytechnics, colleges of education, and colleges of technology, US mass media, those students preparing for Cowbell mathematics competitions, news editors of all major Nigerian newspapers and magazines, news divisions of all TV stations in Nigeria, private, state and national libraries. Of course, the general reader is not left out.

If we are really serious about generating and sustaining interest in mathematics, we might have to start looking for materials outside the basal curriculum and starting with *Mathematical Encounters for the Inquisitive Mind* will not be a step in the wrong direction.

Preparing the Text

Mathematical Encounters For The Inquisitive Mind was typeset initially with a typesetting system called LAT_EX developed by Leslie Lamport. The LAT_EX document uses the 26 letters of the Latin alphabet, the 10 Arabic numerals (0, 1, 2, 3, 4, 5, 6, 7, 8,9), and 32 special characters (which include among others, %, +, $). These comprise a subset of what is known as ASCII *(American Standard Code for Information Interchange)*. ASCII (pronounced ASK—EE) characters are interpreted the same way by all computers: Apple computers, IBM PC, and IBM compatibles (also called clones) which include Dell, Compaq, Panasonic, Zenith, Gateway, and *e*machines etc. LAT_EX creates professional-looking documents using the T_EX typesetting program developed by Dr. Donald Knuth, a math professor at Stanford University in the United States. Dr. Knuth wrote T_EX especially for typesetting mathematical and scientific documents. The capabilities of these programs are amazing. There is no conceivable mathematical symbol that cannot be represented using T_EX: Lines, geometrical shapes, curves, tables, fractions, mathematical symbols etc. are all produced by T_EX commands. With the appropriate commands, the table of contents, chapter, sectional, and sub-sectional headings are all created automatically.

While I take full responsibility for any errors in this text, readers' suggestions for improvement will be appreciated.

Paul Chika Emekwulu
Norman, OK United States of America

Acknowledgments

A project of this magnitude cannot be complete without recognizing people or institutions whose suggestions enhanced the quality of this book or whose actions or so, made this book possible in the first place.

To the United States Immigration & Naturalization Services (a US Govt. agency), for granting me entry visa which made it possible for me to enter the United States to explore opportunities including expressing myself through the power of the pen.

Each chapter of this book is an integral part of a whole. Chapter 8 is a result of an article written by Professor Monte Zerger of Adams State College, Alamosa, Colorado. I thank him for giving me the opportunity to write a rejoinder to his article which was published on page 6 of the January 6, 1996 issue of *The Mathematics Teacher*.

To Mr. Edward Himes, a dedicated American veteran educator who I met as a student at Alvan Ikoku College of Education, Owerri who while engaged in his professional obligation exposed me to the beauty of the Seven Piece Tangram puzzle and National Council of Teachers of Mathematics, I owe much thanks.

To the American high school mathematics teacher whose desire to connect within the curriculum via the internet, I also owe much gratitude. She provided me the opportunity to attempt a contribution to her topic. Every effort was made to contact her but such efforts were all to no avail. My thanks also go to:

The MathPage at www.themathpage.com for the graphs of $y = x^2$ and $y = x^3$ in chapter 3.

Stephen Weierman, adjunct lecturer, City University of New York for his editorial skills and suggestions for improvement.

Jill Britton, a mathematics instructor at Camosun College in Victoria, British Columbia, Canada for permission to use figures in section 3.8.

Chapter 1

Using Line Game to Teach Selected Mathematical Concepts

1.1 Objectives

At the end of the lesson, the students should be able to:

(a) have fun and enjoy the line game.
(b) revise the concepts of percentages, ratio, and fractions.
(d) answer questions based on the line game.

1.2 Introduction

Here in this chapter, we will look at various variations of the game. We will also consider implications of the game for classroom instruction. This will be followed by questions. This chapter will also look at examples of problems on averages, ratio, and percentages. Of special interest is the use of working mean to determine averages.

Number of Players: 2 or More
Materials Needed: Paper, pencil or pen

1.3 How to Play the Game

A certain square array of dots is made on a blank sheet of paper measuring about 8 1/2 by 11 inches. The more players there are, the more the number of dots. The distance between these dots should be the same both vertically and horizontally (see Figure 1). The first player starts the game by joining any two dots either vertically or horizontally. The second player then takes his or her turn. The aim of the player should be to complete a square shape as the dots are being joined by straight lines with the minimum number of moves. But each of the players struggles and makes sure that his opponent does not have the opportunity to form such shapes.

The first player to form such a shape attracts a point to himself or herself. He gives such a shape an individual identification mark e.g. "D" for David and "J" for Jonathan which he or she continues to use till the game is over. After scoring as many points as possible, he or she continues after which the next player takes over.

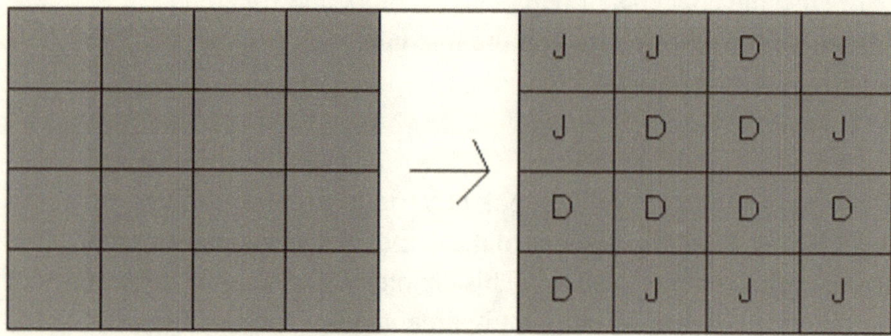

Figure 1: Before and After the Squares are Completed

The game continues till all the dots have been joined. At the end, each player counts his or her square shapes and any of them with the highest number of points is declared the winner. The game can take many variations thus:

1.4 First Variation of the Game

An umpire is appointed. Each shape completed attracts 5 points. After the game, the scores are added up and the player with the highest number of points becomes the winner.

1.5 Second Variation of the Game

Another variation of the game is to have two or three rounds. The points are then added for each of the players. The player with the highest points is declared the winner.

Round	Player 1	Player 2	Total Points
One	20	16	36
Two	18	18	36
Average	19	17	36

Table 1.1: Second Variation of the Game

The last variation is average while the first one teaches addition facts in the elementary grades.

1.6 Implication for Classroom Instruction

This game could be used to reinforce or introduce some mathematical concepts such as percentages, decimal fractions, averages, simple equations, ratio etc. Let us look at a game played by two high school students, David and Jonathan. David and Jonathan are students of Tecumseh High School. A look at figure 1 shows that David had scored 9 out of 25 points while Jonathan scored 16 points. The above situation could generate such questions as:

(a) Who won the game?
(b) What percent of the points did Jonathan score?
(c) What percent of Jonathan's did David score?

(d) How many more points does David need to bring his score up to 40%?

(e) How many points does David need so that Jonathan's score becomes 30% of the total points?

Now, let us solve the fourth question.

SOLUTION: Let the required number of points be x.

But $\frac{\text{Number of points scored}}{\text{Total number of possible points}}{'}\ 100 = 40$

Therefore, $\frac{x+9}{25}{'}\ 100 = 40$

By cross multiplication, $100x + 900 = 25 \times 40$.

Therefore, $100x = (25 \times 40) - 900 = 100$.

Therefore, $x = 100 \div 100 = 1$.

So David needs just one extra point to bring his score up to 40%. We can see from the above discussion that line game can open a lot of other possibilities as a teaching tool. Yes, it can!

Line game can be used as a tool to introduce or teach such concepts as average, ratio, proportion, percentages etc.

Verification of Our Result

David scored 9 out of 25 points. An extra point will increase his points to 10.

Therefore, $\frac{10}{25} \times 100 = \frac{2}{5} \times 100 = \frac{2}{1} \times 20 = 40\%$.

This has been verified.

1.7 Problems on Averages

EXAMPLE 1 For a group of 12 girls, the average mark in an examination is $54\frac{1}{2}\%$. The marks for 11 of them are 55, 56, 52, 50 , 51, 48, 50, 52, 54, 51, 50. (a) Find the marks of the other girl. (b) How many marks should she have scored to bring the average for the group up to 55%?

SOLUTION

(a) Let the other girl's marks $= x\%$.

Total marks for 11 girls

$= 55+56+52+50+51+48+50+52+54+51+50$

$= (50\times 11)+(5+6+2+0+1-2+0+2+4+1+0) = 569$.

$$\text{Average mark} = \frac{\text{Total marks}}{\text{Total number of girls}}$$

Total marks $= 569 + x$ and total number of girls $= 12$.

By substitution therefore, $\frac{569+x}{12} = 54.5$.

(Obeying the principle of equivalent fractions and by cross multiplication.)
$569 + x = 12\times 54.5 = 654$.

From here, $x = 654 - 569 = 85$.%.

1.8 Finding Average Using a Working Mean

We can find the same average above by using the concept of working mean. Working mean involves a process of choosing an arbitrary mean. No matter the mean chosen, regardless whether it is more or less than the actual mean, the total marks would still be the same.

(a) Using 55 as a working mean.

Let the working mean = 55.

Total marks for 11 girls

= (55 × 11) + (55 - 55) + (56 - 55) + (52 - 55) + (50 - 55) + (51 - 55)

+ (48 - 55) + (50 - 55) + (52 - 55) + (54 - 55) + (51 - 55) + (50 - 55)

= 605 + (0 + 1 - 3 - 5 - 4 - 7 - 5 - 3 - 1 - 4 - 5) = 605 - 36 = 569

Total number of girls = 12.

Total marks for 11 girls = 569.

Let the other girl's marks = x%.

Therefore, total marks for 12 girls = (569 + x).

But average marks for 12 girls = $\dfrac{\text{Total marks for 12 girls}}{\text{Total number of girls}} = \dfrac{569+x}{12}$

Therefore, $\dfrac{569+x}{12} = 54.5$

By cross multiplication, $12 \times 54.5 = 569 + x$.

$654 = 569 + x$.

From here, $x = 654 - 569 = 85$.

The 12th girl scored 85%.

Using 60 as a working mean

Let the working mean = 60.

Total marks for 11 girls

= (60 × 11) + (55 - 60) + (56 - 60) + (52 - 60) + (50 - 60) + (51 - 60)

+ (48 - 60) + (50 - 60) + (52 - 60) + (54 - 60) + (51 - 60) + (50 - 60)

= 660 + (-5 - 4 - 8 - 10 - 9 - 12 - 10 - 8 - 6 - 9 - 10) = 660 - 91 = 569.

Total number of girls = 12.

Total marks for 11 girls = 569.

Let the other girl's marks = x%.

Therefore, total marks for 12 girls = (569 + x).

But average marks for 12 girls = $\frac{\text{Total marks for 12 girls}}{\text{Total number of girls}} = \frac{569 + x}{12}$

Therefore, = $\frac{569 + x}{12} = 54.5$

By cross multiplication, $12 \times 54.5 = 569 + x$.

$654 = 569 + x$.

From here, $x = 654 - 569 = 85$.

The 12th girl scored 85%.

(b) Let her score = y%.

Total marks for 12 girls = 569 + y.

But marks for 12 girls = $\frac{\text{Total marks for 12girls}}{\text{Total number of marks}} = \frac{569+y}{12}$.

Therefore, $\frac{569+y}{12} = 55$

By cross multiplication, $12 \times 55 = 569 + y$.

$660 = 569 + y$

From here, $y = 660 - 569 = 91$.

Therefore, the 12th girl should have scored 91% to bring the average of the group up to 55%.

1.9 Problems on Ratios

EXAMPLE 1 If 135: x = 3: 5, what is x? **(S.C./GCE., 1978)**

SOLUTION 135: x = 3: 5.

$135 : x = 3 : 5 \Longleftrightarrow \frac{135}{x} = \frac{3}{5}$.

By principle of equivalent fractions,

$3x = 135 \cdot 5$.

Dividing both sides by 3 we have:

$\frac{5 \times 135}{3} = \frac{3x}{3}$

Therefore, $x = 5 \cdot 45 = 225$.

EXAMPLE 2 (b) Kofi, Afua, and Braima shared a certain amount of money in the ratio of $4\frac{1}{2} : 6 : 10\frac{1}{2}$ respectively. If Braima has $N1.20$ more than Afua, how much does Kofi receive? **(S.C./G.C.E 1982)**

SOLUTION Let the amount shared be x kobo.

Ratio in which the money was shared

$$4\frac{1}{2}:6:10\frac{1}{2}=\frac{9}{2}{'}\ 2:6{'}\ 2:\frac{21}{2}{'}\ 2=9:12:21=3:4:7$$

Total ratio $=3+4+7=14$.

Total money shared $=14x$.

Afua's share $=\frac{4}{14}x=\frac{2}{7}x$

Braima's share $=\frac{7}{14}x=\frac{1}{2}x$.

But Braima has $N1.20$ more than Afua.

Therefore, $\frac{1}{2}x-\frac{2}{7}x=120$.

Clearing fractions, we have:

$7x-4x=1680 \Longleftrightarrow 3x=1680$.

Therefore, $x=1680\div 3=560$.

Therefore, Kofi received $\frac{3}{14}.560$

$=\frac{3\times 40}{1}\times\frac{14}{14}=N1.20$. **Answer:** Kofi's share $=N1.20$.

EXAMPLE 3 A car traveled from Lagos to Ipeku with an average speed of 36 mph for the first 21 miles and 40 mph for the last 30 miles of the journey. What is the average speed for the whole journey? **(S.C./GCE 1980)**

SOLUTION **First part of the journey**

Distance for the first part of the journey $= 21$ miles.

Average speed for this journey $= 36$ mph.

Therefore, time taken to cover this distance

$$= \frac{21}{36} = \frac{3}{3}, \frac{7}{12} = \frac{7}{12}$$

Second part of the journey

Distance covered $= 30$ miles.

Speed of car $= 40$ mph.

Therefore, time taken $= \frac{30}{40} = \frac{3 \times 10}{4 \times 10} = \frac{3}{4}$ hr.

Total time taken $= \frac{7}{12} + \frac{3}{4} = \frac{7}{12} + \frac{9}{12} = \frac{16}{12}$.

But average speed $= \frac{\text{Total distance covered}}{\text{Total time taken}} = \frac{21+30}{\frac{16}{12}}$.

$$= \frac{51}{16} \times \frac{12}{1} = \frac{51 \times 4 \times 3}{4 \times 4} = \frac{51 \times 3}{4} = \frac{153}{4} = 38.25 \text{ mph.}$$

1.10 Problems on Percentages

EXAMPLE 1 A trader sells rice in 50kg—bags. 30 bags of rice costing N40 per bag are mixed with 50 bags of rice of another kind costing *N*35.00 per bag. If he sells the mixture at a gain of 20%, at what price does he sell a bag?
(S.C./G.C.E, 1978)

SOLUTION Total number of bags $= 80$.

Cost of 50 bags @ N35.00 per bag = N35 × 30 = N1,750.00..
Cost of 30 bags @ N40.00 per bag = N40 × 30 = N1,200.00.

Total cost for 80 bags $= N2950.00$.

This amount is represented by 100%.

Therefore, $100\% = N2,950.00$.

$120\% = ? = \frac{120}{100} \times 2950 = \frac{6 \times 20 \times 5 \times 590}{20 \times 5}$

Selling price $= N(6 \times 590)$ for 80 bags.

Selling price per bag $= \frac{6 \times 590}{80} = \frac{2 \times 3 \times 10 \times 59}{2 \times 4 \times 10} = \frac{177}{4} = 44.25$.

Therefore, cost of 1 bag of mixture $= N44.25$.

EXAMPLE 2 The combined mass of a matching nut and a bolt is 100 gm. A dealer buys a quantity of such matching pairs from Liberia at a cost (including freight and other expenses) of \$4,800.00 per tonne. For how much must he sell a pair in Nigeria to make a profit of 25% given that 1.00 Naira = \$2.40 **(S.C./GCE 1980).**

SOLUTION Combined mass of a matching nut and a bolt $= 100$ gm.

Cost of a certain number of such matching pair $= \$4,800.00$.

Then, $\frac{4800 \times 100}{240} = 2000.00$ Naira.

Therefore, the cost of the matching pairs in Nigerian Naira = 2000

Therefore, number of pairs $= \frac{1000 \times 1000}{100} = 10000$ (1000 gm = 1 kg)

But every 1000 kg (1 tonne) cost 2000. 00 naira.

Cost of a pair $= \frac{2000}{10000} = 0.20$ naira.

Therefore, $100\% = 20$ kobo.

$$125 = \frac{125}{100} \times \frac{20}{1} = \frac{5 \times 25}{5 \times 20} \times \frac{4 \times 5}{1}$$

$$= \frac{5}{5} \times \frac{5\,4}{5\,4} \times \frac{25}{1} = 25$$

The profit is 25 kobo.

EXAMPLE 3 A trader bought an article for \$4.50 and sold it for a profit of 20% to Mr. B who made a profit of 30% by selling it to a customer. What price did the customer pay for the article ?

SOLUTION 20% of \$4.50 $= \frac{20}{100} \times 450 = 90$ cents.

Therefore, B's cost price = $\$(4.50 + 0.90) = \$5.40$.

B's profit $= \frac{30}{100} \times \frac{540}{1} = \$1.62$.

Therefore, B's selling price $= \$(5.40 + 1.62) = \$7.02$

and this is the price paid by the customer.

1.9 A Search for Counter Examples

Can you identify two equal ratios a:b and x:y such that:

$$\frac{a}{b} \neq \frac{x}{y}?$$

$$or\ bx \neq ay\ or\ \frac{a}{x} \neq \frac{b}{y}\ or\ \frac{x}{a} \neq \frac{y}{b}?$$

Chapter 2

Using Your Intuition for Self-Empowerment

2.1 Objectives

At the end of the lesson, the students should be able to:

(a) Realize that ultimate reality is beyond the five physical senses.
(b) Learn from examples of the power of intuition from true life experiences.
(c) Realize that we all have or can develop our intuitive faculties.
(d) State some properties of Fibonacci and Lucas numbers.

2.2 Introduction

Intuition, through dreams or other psychic experiences is at the core of many breakthroughs in history. Shouldn't we learn to trust it more? Here is an advice from an author who was inspired to write a book by a dream experience.

In wartime England, as the Prime Minister heads for an important meeting something tells him not to get into the waiting car. Moments later, the vehicle exploded.

When Archimedes ran out of his bathtub naked shouting, "Eureka! Eureka!" (I have found it! I have found it!) he was not crazy, he was not possessed by the evil spirit. It was as a result of intuition. It was as a result of inspiration. Archimedes today is remembered for his famous principle in Physics. Archimedes discovered how to calculate specific gravity by measuring displacement of equal volume of water.

In 1960, a Chicago businessman ignored the advice of financial experts. Instead he trusted his hunches and purchased a small chain of hamburger stands.

Mozart claimed that he received his inspiration from within. Socrates said he was guided by his inner voice and so did Einstein, Edison, Marconi, Henry Ford, Luther, Madame Curie, and other Nobel Laureates.

2.3 The Neglected Power

Einstein's Theory of Relativity, Churchill's hesitation to enter the waiting car, Archimedes' experience in his bath tub and Ray Kroc's McDonald's are all examples of the power of intuition. Intuitive experience has been described with many terms: inspiration, sixth sense, hunches, and insight, to name a few. Many of the greatest world philosophers, artists, writers, musicians, and scientists such as Einstein, Mozart, and Beethoven have owed their accomplishments to intuition.

Articles on this subject have been carried in such publications as Entrepreneur Magazine, Fortune 500, Forbes Magazine, and the Wall Street Journal. Yet, since intuition is beyond scientific explanation, it is both the most powerful and the most neglected of human mental faculties. All along, intuition has been part of the instruction of most mystical schools or organizations.

Their emphasis is on the development of the higher self. Unfortunately, people are busy looking for empirical evidence of intuition. The end result of this search is disappointment. Yes, disappointment, unless the search is a genuine one and not out of skepticism. According to *Perspective* on the ABC New TV Program, such genuine researches are now currently in progress at the University of Edinburgh in Scotland and at the University of Nevada in the United States.

Intuitive ability is a potential, but some of us make a conscious effort to either block, repress or deny its existence and consequent manifestation.

2.4 Intuitive Dreaming

Dr. Frederick Banting, the Canadian physician, discovered, while in a dream state, the basis of insulin. Elias Howe, the inventor of the sewing machine, worked for years for its design, yet one thing always remained for him to achieve his goal. One thing! One night he had a dream he had been captured by savages who were directing spears at him, and he noticed that at the tip of each spear was a hole. Howe woke from his dream with the solution: put the hole at the tip of the needle. That key unlocked the invention of the sewing machine.

Noble prize winner Dr. James Watson had been working trying to decipher the molecular structure of DNA, for many years, but all to no avail. One night while in a dream, he saw two snakes coiling around each other. Instantly, he woke up exclaiming, "I wonder if that's it? I wonder if DNA is a double helix twining around itself?" This led him to understand the genetic coding for which he was awarded a Noble Prize.

2.5 An Exceptional Dream Experience

Intuition is real, not a fantasy. I believe in intuition. Why? Because it exists, and proofs of its existence abound. I even wrote one of my books, *Fibonacci Numbers For Research Mathematicians & AI Applications*, inspired by a dream experience.

A set of numbers like 0, 2, 4, 6, 8, 10 (geometric series), increases uniformly in twos. Therefore, it exhibits a pattern. A set of numbers like the one above is also called a sequence or series. A sequence can take other forms other than the example given above. Having offered this explanation, let's go back to my dream experience.

In the wee hours of the night of March 30, 1993, I had a dream where I saw a partial floral representation of the Fibonacci sequence (see Figure 1). For some who are not familiar with the Fibonacci sequence, it runs like this:

1, 1, 2, 3, 5, 8, 13, 21, 34, 55, 89, 144, 233, 377, 610, 987, 1597, 2584, $x, y, x + y \ldots$

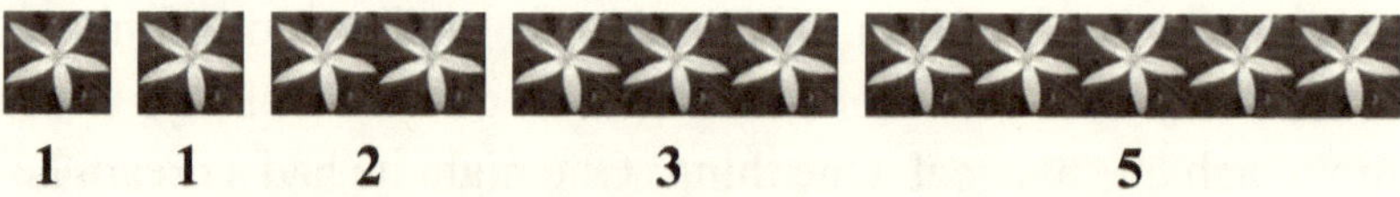

1 1 2 3 5

Figure 2.1: Floral Representation of a Subsect of the Fibonacci Sequence

This sequence was discovered by an Italian mathematician named Leonardo Pisano, and each of the numbers of the sequence is called a Fibonacci number. One outstanding property of this sequence is that the next term is a sum of the two preceding terms, e.g.

$1 + 1 = 2$, $1 + 2 = 3$, $2+3 = 5$, $3+5 = 8$, $5+8 = 13$, $8+13 = 21$, $13 + 21 = 34$

The following morning I took up a piece of paper and a pen. Initially, I thought it would have been an article of about few pages. I was wrong. As days passed by, the more I played with those numbers, the more I discovered properties of the sequence. The properties are as exciting as they are interesting. So I ended up with about 150 pages of material out of sheer inspiration.

Numbers of the Fibonacci sequence have many properties. Fibonacci numbers appear in nature. They have applications in arts, architecture, artificial intelligence, and other fields. Anyway, discussing these and other properties is not the objective of this chapter. What was exciting to me in this dream experience was the transformation of consciousness.

During and after the dream experience, I was extremely calm, peaceful, and serene. I am short of words to describe my actual experience, but have not in my forty-something years of existence on this planet, attained that level of consciousness. It was amazing!

2.6 Learning to Listen to Your Inner Voice

One of our weaknesses as human beings is our continued tendency to believe that reality can only be perceived through the five mundane senses of sight, smell, touch, taste and hearing. To some people, it is a matter of seeing it before they believe it. They want to smell it. They want to touch it. They want to taste it. They want to hear it.

Such thoughts and ideas in more or less logical forms, need be captured and written down, otherwise, they could disappear. Even if they are not entirely lost, they might lose their original flavor.

Therefore, in order to maintain the present level of civilization, and still further human progress, we have to maintain open minds to these experiences. We all have an inner voice. We just have to improve our listening skills.

What more can we really say about Fibonacci numbers?

Quite a lot can be said about Fibonacci numbers.

2.7 Amazing Properties of Fibonacci and Lucas Numbers

Fibonacci numbers as we have already learnt are numbers of the form:

1, 1, 2, 3, 5, 8, 13, 21, 34, 55, 89, 144, 233, 377, 610, 987, 1597, 2584, 4181, x, y, $x+y$ . . .

while Lucas numbers are numbers of the form:

1, 3, 4, 7, 11, 18, 29, 47, 76, 123, 199, 322 . . .

The innocent-looking Fibonacci sequence has so many intriguing and interesting properties. The Fibonacci sequence has many untold properties and trying to provide all of them here is not just possible. Moreover, despite the wide interest in the mathematical community, and conference proceedings on Fibonacci numbers and its applications, the sequence is still an unexplored territory.

Property 1: Let $\{a_n\} = 1, 3, 4, 7, 11, 18, 29, 47, 76, 123, 199 \ldots$

$$a_1 = 1$$
$$a_2 = a_0 + a_1$$
$$a_3 = a_1 + a_2$$
$$a_4 = a_2 + a_3$$
$$a_5 = a_3 + a_4$$

n	n-2	n - 1
2	0	1
3	1	2
4	2	3
5	3	4

Table 1: In search of a pattern

The following are true about a_n.

Generally, $S_n = a_{n+2} - 3, n \geq 2$.

Figure 2.2: Edouard Lucas, 1842-1891
Source: Wikipedia, the free Encyclopedia

2.8 Difference between First n Sums of Lucas Numbers

$S_1 - S_0 = 1 - 0 = 1$
$S_2 - S_1 = 4 - 1 = 3$
$S_3 - S_2 = 8 - 4 = 4$
$S_4 - S_3 = 15 - 8 = 7$
$S_5 - S_4 = 26 - 15 = 11$
$S_6 - S_5 = 44 - 26 = 18$
$S_7 - S_6 = 73 - 44 = 29$
$S_8 - S_7 = 120 - 73 = 47$
$S_9 - S_8 = 196 - 120 = 76$
$S_{10} - S_9 = 319 - 196 = 123$

Generally, $a_n = S_n - S_{n-1}, n \geq 1$

Validity of Result

By substitution for $S_n - S_{n-1}$ we have:

$$S_n - S_{n-1} = (a_{n+2} - 3) - S_{n-1}$$

$$= (a_{n+2} - 3) - a_{n-1+2} - 3$$

$$= (a_{n+2} - 3) - (a_{n+1} - 3)$$

$$= (a_{n+2} - a_{n+1}) - 3 + 3$$

$$(a_{n+2} - a_{n+1}) + 0 = a_n$$

Therefore, $a_n = S_n - S_{n-1}, n \geq 1$.

Property 2: The product of first two successive n terms of the Fibonacci sequence (u_n), results to a sequence (T_n) whose consecutive terms add up to form a subset of the Fibonacci sequence with even subscripts.

$1 \times 1 = 1$	$1 \times 2 = 2$
$2 \times 3 = 6$	$3 \times 5 = 15$
$5 \times 8 = 40$	$8 \times 13 = 104$
$13 \times 21 = 273$	$21 \times 34 = 714$
$34 \times 55 = 1870$	$55 \times 89 = 4895$

By multiplying the numbers as we did above, we generate an entire new set of numbers. These are not entirely Fibonacci Numbers. Not yet! The level of excitement at this point depends on the behavior of this new set of numbers. Let us call this set T_n.

Now, take a look!

$T_1 = (u_1)(u_2)$	$T_7 = (u_7)(u_8)$
$T_2 = (u_2)(u_3)$	$T_8 = (u_8)(u_9)$
$T_3 = (u_3)(u_4)$	$T_9 = (u_9)(u_{10})$
$T_4 = (u_4)(u_5)$	$T_{10} = (u_{10})(u_{11})$
$T_5 = (u_5)(u_6)$	$T_{11} = (u_{11})(u_{12})$
$T_6 = (u_6)(u_7)$	$T_{12} = (u_{12})(u_{13})$

Table 2: Finding a Relationship Between n and k

Each of these subtraction facts can be represented as $n - 1 = k$.

Therefore, generally, $T_n = (u_n)(u_{n+1})$,.

Also, from table 3, $n - k = 1$.
From here, $k = n - 1$.

n	$n-1$	k	n	$n-1$	k
2	1	1	7	6	6
3	2	2	8	7	7
4	3	3	9	8	8
5	4	4	10	9	9
6	5	5	11	10	10

Table 3: k = n - 1, n-k = 1

2.9 Difference between Consecutive Terms

Let5 us investigate the differences between consecutive terms of our new sequence.

$T_2 - T_1 = 2 - 1 = 1 = 1^2$	$T_7 - T_6 = 273 - 104 = 169 = 13^2$
$T_3 - T_2 = 6 - 2 = 4 = 2^2$	$T_8 - T_7 = 714 - 273 = 441 = 21^2$
$T_4 - T_3 = 15 - 6 = 9 = 3^2$	$T_9 - T_8 = 1870 - 714 = 1156 = 34^2$
$T_5 - T_4 = 40 - 15 = 25 = 5^2$	$T_{10} - T_9 = 4895 - 1870 = 3025 = 55^2$
$T_6 - T_5 = 104 - 40 = 64 = 8^2$	$T_{11} - T_{10} = 12816 - 4895 = 7921 = 89^2$

A pattern is here with us. The difference is turning out to be square numbers.

Yes, the difference seems to be square numbers. Of course, you should have noticed that the square roots of these square numbers are also Fibonacci numbers.

How interesting!

Again, take a look at the sequence, $T_{n.}$
Doing so we have:

$1+2=3$	$104+273 = 377$
$2+6=8$	$273+714 = 987$
$6+15=21$	$714+1870 = 2584$
$15+40 = 55$	$1870+4895 = 6765$
$40+104 = 144$	$4895+12816=17711$

In other words, the following are true and let us also extract some of the subscripts and present them as shown below:

$n-1$	n	k		$n-1$	n	k
1	2	4		5	6	12
2	3	6		6	7	14
3	4	8		7	8	16
4	5	10		8	9	18

$k = n+(n-1)+1 = 2n$.

Therefore, $T_{n-1}+T_n = u_{2n}$, where $n \geq 2$

The sum of first n terms of T_n which are derived by multiplying consecutive terms of the Fibonacci sequence are themselves Fibonacci Numbers.

Summarily, the product of first n terms of the Fibonacci sequence results to a new sequence T_n, whose preceding terms add up to form a subset of the Fibonacci sequence with even subscripts.

Verifying Our Result

If $T_n + T_{n-1}$ is a Fibonacci number, then it can be expressed as a number, u_k

Now, $T_{n-1} + T_n = (u_n)(u_{n-1}) + (u_{n+1})(u_n)$

$= u_n(u_{n-1} + u_{n+1}) = (u_n)(a_n) = u_{2n}$.

Any Fibonacci number u_k whose subscript is even can be represented as a product of two numbers, a Lucas number a_n and a Fibonacci number u_n whose individual subscript is the sum of the subscripts of a_n and u_n or half subscript of u_k.

Therefore, $T_n + T_{n-1}$ is a Fibonacci number.

Property 3: The difference between squares of first n terms of the Fibonacci sequence (u_n) is equal to a sequence (g_n) whose sum of first n terms is equal to a subsect of the Fibonacci sequence with even subscripts.

Let us again consider the Fibonacci sequence. Each term of the sequence as we have already noticed is u_n. Our immediate objective is to find $(u_n)^2$ and $(u_{n+1})^2 - (u_n)^2$.

n	(u_n)	$(u_n)^2$	$(u_{n+1})^2$	$(u_{n+1})^2$ - $(u_n)^2$
1	1	1	1	0
2	1	1	4	4-1 = 3
3	2	4	9	9-4 = 5
4	3	9	25	25-9 = 16
5	5	25	64	64-25 = 39
6	8	64	169	169-64 = 105
7	13	169	441	441-169 = 272

$g_n = \{0, 3, 5, 16, 39, 105, 272, 715, 1869, 4896, 12815, 33553, \cdots\}$.

The first n terms of the sequence add up to Fibonacci Numbers. Take a look!

$g_1 + g_2 = 0 + 3 = 3 = u_4$

$g_2 + g_3 = 3 + 5 = 8 = u_6$

$g_3 + g_4 = 5 + 16 = 21 = u_8$

$g_4 + g_5 = 16 + 39 = 55 = u_{10}$

$g_5 + g_6 = 39 + 105 = 144 = u_{12}$

$g_6 + g_7 = 272 + 105 = 377 = u_{14}$

$g_7 + g_8 = 715 + 272 = 987 = u_{16}$

$g_8 + g_9 = 1869 + 715 = 2584 = u_{18}$

$n + (n+1) + 1 = k$

Therefore, k = (n + 1) + (n + 1) = 2n+2.

Therefore, generally, $g_n + g_{n+1} = u_{2n+1}$.

Generally, if $g_n = (u_{n+1})^2$, then $g_n + g_{n+1} = u_{2n+2}$

where k is a member of the set of Natural numbers.

2.10 Proof that $g_n + g_{n-1}$, n ≥ 2, is a Fibonacci number with an even subscript

Proof

If $g_n + g_{n-1}$ is a Fibonacci number, then it can be expressed as a number u_k

where k is the subscript.

By definition, $g_n = (u_{n+1})^2 - (u_n)^2$

and $g_{n-1} = (u_{n-1+1})^2 - (u_{n-1})^2 = (u_n)^2 - (u_{n-1})^2$.

By substitution, $g_n + g_{n-1} = (u_{n+1})^2 - (u_n)^2 + [(u_n)^2 - (u_{n-1})^2]$

$= (u_{n+1})^2 - (u_{n-1})^2$

$= (u_{n+1} + u_{n-1})(u_{n+1} - u_{n-1}) = (a_n)(u_n) = u_{2n}$.

Property 4:

Let (i) u_n be the nth Fibonacci number.
(ii) Number of digits in $u_n = n_k$.
(iii) $u_n + n_k = R_n$.

Figure 2.3 on page 51 shows the sum of a Fibonacci number u_n

and the sum of digits in u_n while figure 5 shows the difference between successive terms of R_n i.e. $R_{n+1} - R_n$.

There is a relationship between $u_n + n_k$ on one hand and $u_{n+1} + n_{k+1}$ on the other hand. This relationship is a property and could be stated as follows:

The sum of number of digits of first *n* terms of the Fibonacci sequence and consecutive terms of the entire Fibonacci sequence forms a sequence, R_n whose difference of consecutive terms is equal to a Fibonacci number with the exception of transition points.

Transition points as used here are defined as those points on the Fibonacci sequence where the number of digits increases from *n* to $n+1$. A transition point, $R_{n+1} - R_n$ is not a Fibonacci number and using a method of proof called *proof by contradiction*, we can show that this is true.

$R_{n+1} - R_n$

$2 - 2 = 0$

$3 - 2 = 1$

$4 - 3 = 1$

$6 - 4 = 2$

Proof

Let u_n = n[th] Fibonacci number, while n_k= number of digits in u_n.

Each subtraction fact in figure 4 can be represented by $R_{n+1} - R_n$.

By substitution, in $R_{n+1} - R_n$ we have:

$$R_{n+1} - R_n = (n_{k+1} + u_{k+1}) - (n_k + u_k))$$

$$= (n_{k+1} - n_k) + (u_{k+1} - u_k)$$

$$= 1 + u_{k+1} - u_k$$

$$= (u_{k+1} - u_k) + 1$$

But k = n.

Then by substitution, we have:
$(u_{k+1} - u_k) + 1 = (u_{n+1} - u_n) + 1$.

The quantity $u_{n+1} - u_n$ is a Fibonacci number but $(u_{n+1} - u_n) + 1$ is a number 1 more than a Fibonacci number. Therefore, at transition points, the sum of number of digits in a Fibonacci number and the corresponding Fibonacci number is not a Fibonacci number. We can prove this alternatively.

Fibonacci numbers (u_n)	**Number of Digits (n_k)**	$R_n = u_n + n_k$
1	1	2
1	1	2
2	1	3

3	1	4
5	1	6
8	1	9
13	2	15
21	2	23
34	2	36
55	2	57
89	2	91
144	3	147

Figure 2.3: $R_n = u_n + n_k$

Transition Point
2-2=0
3-2=1
4-3=1
6-4=2
9-6=3
15-9=6
23-15=8
36-23=13
57-36=21
91-57=34
147-91=56

Figure 2.4: $\mathbf{R_{n+1} - R_n}$

Proof

If at a transition point, the sum of a Fibonacci number and its corresponding subscript is a Fibonacci number, then

$$R_{n+1} - R_n = u_{n+1} - u_n$$

$$R_{n+1} - R_n = (n_{k+1}) - (n_k + u_k)(n_{k+1} - n_k) + (u_{k+1} - u_k)$$

$$= 1 + (u_{k+1} - u_k) = (u_{k+1} - u_k) + 1..$$

But k = n.
By substitution, we have:

$$R_{n+1} - R_n = (u_{n+1} - u_n) + 1$$

and this is not a Fibonacci number but a number 1 more than a Fibonacci number.

REMARK: The above proof is called *proof by contradiction*. In proving by contradiction, we assume that a false statement is true, and then end up proving it to be false.

2.11 Practical Examples of Intuition

A great many men of science, poets, philosophers, musicians, writers and others have declared that they received important ideas and suggestions in dreams. The following are some of them.

Dmitri Bendeleyev—Periodic Table of Elements Discovered in a Dream

The periodic table of chemical elements was created as a result of a dream experienced by Dmitri Mendeleyev, a professor of Chemistry at St. Petersburg, Russia. This was in 1869.

Dmitri Mendeleyev was born on February 8, 1834 in Tobolsk in western Siberia and youngest of 14 or 17 children. In 1869 he was puzzling over the problem of the chemical elements. This time there were 63 different known elements ranging from copper and gold to rubidium. Every one of these elements consisted of different atoms and that the atom of each element had their own unique properties. However, it had been found that some elements do possess similar properties. This allows for easier classification in groups.

Figure 2.5: Dmitri Mendeleyev (1834-1907)
Source: Wikipedia, the free Encyclopedia

The atoms making up the different elements were known to have different atomic weights. The lightest of them is hydrogen with an atomic weight of 1. The heaviest known element, lead was thought to have an atomic weight of 207. This meant that the elements could be listed linearly—according to ascending atomic weights or alternatively, they could be classified in groups with similar properties. Several scientists had begun to suspect that there exists a link between these two methods of classification.

Yes, the elements had different weights and yes also, they had different properties. They could be numbered and they could be grouped. There just had to be a link between the two pattern forms. At the age of 32, Mendeleyev was appointed a professor of general chemistry at St. Petersburg University and was well known for his encyclopedic knowledge of the elements. He was looking for a pattern that linked the chemical properties of the elements.

At regular numerical intervals, certain similar properties seemed to repeat in the elements. a few of the intervals began with a certain regularity, but then disappears. Despite this inconsistency, Mendeleyev knew he was in the brink of major scientific breakthrough. He was convinced of this. He became exhausted and almost immediately, he fell asleep and had a dream.

In Mendeleyev's own words: I saw in a dream a table where all the elements fell in place. Awakening I immediately wrote it down on a piece of paper.

Periodic table of the elements has found applications in Chemistry, Physics, and Engineering especially chemical engineering.

The Riddle of Benzene Molecule - Friedrich A. von Kekule

Friedrich A. von Kekule, a professor of Chemistry at Ghent University, Belgium, had been struggling to solve the structural riddle of the benzene molecule. His dream discovered insight was presented to a scientific convention in 1890.

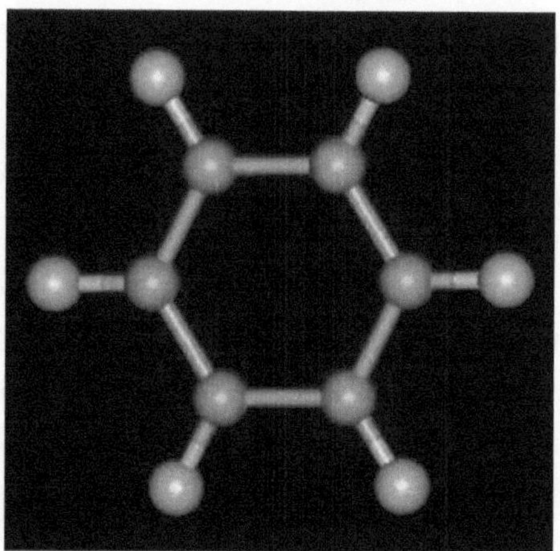

Figure 2.6: Benzene Molecule

"I fell into a reverie, and lo, the atoms were gamboling before my eyes! Whenever, hitherto, these diminutive beings had appeared to me, they had always been in motion but up to that time, I had never been able to discern the nature of their motion. Now, however, I saw no, frequently, two smaller atoms united to form a pair, how a larger one embraced the two smaller ones; how still larger ones kept hold of three or even four of the smaller; whilst the whole kept whirling in a giddy dance. I saw how the larger ones formed a chain, dragging the smaller ones after them, but only at the ends of the chain. The cry of the conductor: Clapham Road, awakened the night in putting on paper at least sketches of these dream forms. This was the origin of the Structural Theory.'

Dr. Frederick Banting and the Insulin

Dr. Fredrick Banting, a Canadian physician discovered, while in a dream state, the basis of insulin.

Figure 2.7: Dr. Frederick Banting (1891-1941)

Elias Howe and the Sewing Machine

Elias Howe, the inventor of the sewing machine, worked for years for his design, yet one thing always remained for him to achieve his goal. One thing! One night he dreamt he had been captured by savages who were pointing spears at him, and he noticed that at the tip of each spear was a hole. Howe woke from his dream with the solution: put the hole at the tip of the needle. That key unlocked the invention of the sewing machine. Amazing!

Figure 2.8: Elias Howe, inventor of the Sewing Machine

Madame C. J. Walker - From Dream to Millionaire

Madame C.J. Walker (1867-1919), according to the Guinness Book of World Records, is the first female American self-made millionaire. Madame Walker was the founder of a successful African-American cosmetic company that turned her into a millionaire. Walker had a scalp infection that resulted in much hair loss. She resorted to experimenting with patented medicines and hair care products. Eventually, she went through a dream experience that solved her problem. "He answered my prayer, for one night I had a dream, and in that dream, a big black man appeared to me and told me what to mix up in my hair. Some of the remedy was grown in Africa, but I sent for it, mixed it, put it on my scalp, and in a few weeks my hair was coming in faster that it had ever fallen out. I tried it on my friends, it helped them. I made up my mind to begin to sell it."

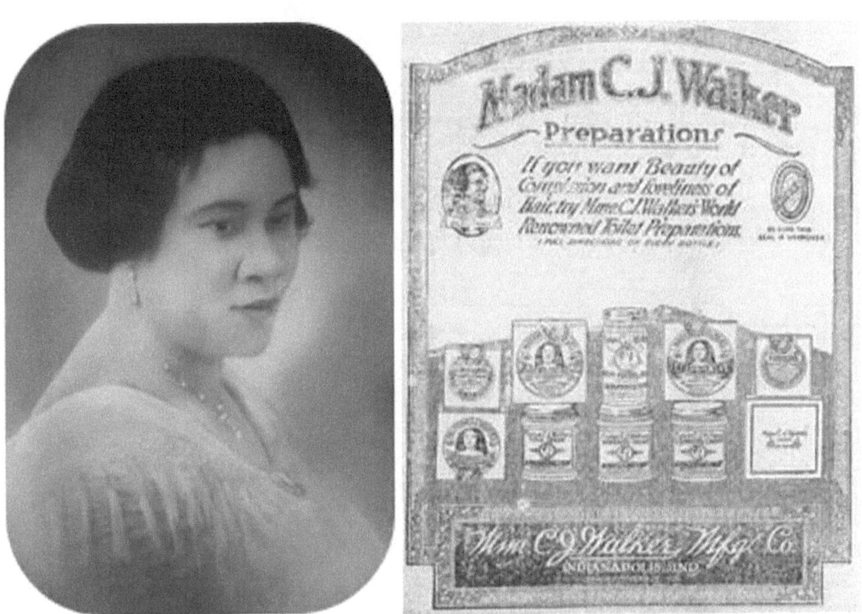

Figure 2.9: Madame C.J. Walker (1867-1919)

Dr. James Watson - Structure of DNA Revealed in a Dream

Dr. James Watson a noble prize-winner, had been working trying to decipher the molecular structure of DNA, for donkey years, but all to no avail. One night while in a dream, he saw two snakes coiling around each other. Instantly, he woke up exclaiming, "I wonder if that's it? I wonder if DNA is a double helix twining around itself"? This led him to understand the genetic coding for which he was awarded a Nobel Prize.

Figure 2.10: Dr. James Watson

Jack Nicklaus Found a New Golf Swing in a Dream

Jack Nicklaus found a new golf swing in a dream which he attributed to improving his golf game. In 1964 Nicklaus was having a bad slump. He was doing badly. Suddenly, his scores went up dramatically and he reported the following:

"Wednesday night I had a dream and it was about my golf swing. I was hitting them pretty good in the dream and all at once I realized I wasn't holding the club the way I've actually been holding it lately. I've been having trouble collapsing my right arm taking the club head away from the ball, but I was doing it perfectly in my sleep. So when I came to the course yesterday morning I tried it in the way I did in my dream and it worked. I shot a sixty-eight yesterday and a sixty-five today."

Figure 2.11: Jack Nicklaus

Dream Gives Birth to a Musical Instrument

Ernst Chladni had been working for a musical instrument which would possess a special sound. In 1789, he awoke from a dream with the complete detail of how the instrument should look. His dream eventually gave birth to the euphonium, a brass instrument like the tuba but having a more mellow tone.

Figure 2.12 : Ernst Chladni

Ray Krock McDonald—Chain of Restaurants Built on Intuition

Raymond "Ray" Allbert Krock McDonald (1902-1984) was a Chicago businessman who, in 1960, ignored the advice of financial experts. Indeed he trusted his hunches and purchased a small chain of hamburger stands.

Figure 2.13: BigMac Combo
Source: Wikipedia: free encyclopedia

Abraham Lincoln Dreamed of His Assassination

President Abraham Lincoln gave an account of the following dream to his wife a few days before his assassination. "About ten days ago, I retired very late. I had been up waiting for important dispatches from the front. I could not have been long in bed when I fell into a slumber, for I was weary. I soon began to dream. There seemed to be a death-like stillness about me. Then I heard subdued sobs, as if a number of people were weeping. I thought I left me be and wandered downstairs. There the silence was broken by the same pitiful sobbing but the mourners were invisible. I went from room to room; no living person was in sight, but the same mournful sounds of distress met me as I passed along. It was light in all the rooms; every object was familiar to me; but where were all the people who were grieving as if there hearts would break?

I was puzzled and alarmed. What could be the meaning of all this? Determined to find the cause of a state of things so mysterious and so shocking, I kept on until I arrived at the East room, which I entered. There I met with a sickening surprise. Before me was a catafalque, on which rested a corpse wrapped in funeral vestments. Around it were stationed soldiers who were acting as guards; and there was a throng of people, some gazing mournfully upon the corpse whose face was covered, others weeping pitifully.

Figure 2.14: Assassination of President Abraham Lincoln

'Who is dead in the White House?' I demanded of one of the soldiers. 'The President' was his answer; 'he was killed by an assassin.' Then came a loud burst of grief from the crowd, which awoke me from my dream."

Mathematical Genius & Dreamer—Srinivasa Ramanujan

Srinivasa Ramanujan (1887-1920), "a major contributor to modern number theory who claimed that he received many of his mathematical concepts from the Indian goddess Namagiri. In his dreams, he was presented with mathematical formulae which he would verify on awakening. This pattern of receiving mathematical formulae "repeated itself throughout his life."

$$\frac{1}{\text{p}} = \frac{2\sqrt{2}}{980!}\overset{\text{¥}}{\underset{k=0}{\text{å}}}\frac{(4k)!(1103+26396)}{(k!)^{k}396^{4k}}$$

Ramanujan describes one of his dreams of mathematical discovery

Figure 2.15: Srinivasa Ramanujan (1887-1920)

"While asleep I had an unusual experience. There was a red screen formed by flowing blood as it were. I was observing it. Suddenly a hand began to write on the screen. I became all attention. That hand wrote a number of results in elliptic integrals. They stock in my mind. As soon as I woke up, I committed them to writing . . ."

Archimedes—Specific Gravity Discovered in a Dream

Archimedes When Archimedes ran out of his bathtub naked shouting, "Eureka, Eureka," (I have found it!, I have found it!) he was not crazy, he was not possessed by the evil spirit. It was as a result of intuition. Archimedes discovered how to calculate specific gravity by measuring displacement of equal volume of water.

Figure 2.16: Archimedes of Syracuse
Source: http://inventors.about.com

J. Schelyer—Synthetic Language Discovered in a Dream

J. Schelyer In 1879, a new synthetic language was created in a dream by J. Schleyer, the German linguist. In thirty years of study, Schelyer has mastered more than fifty languages and was seeking a way to express the common patterns that existed between their grammatical structures. None of his efforts yielded a successful result. In a dream, the necessary letters, forms, and processes appeared "in an ordered array" and his new language was created for auxiliary use in international communication.

Some have drawn inspiration from their inner voices as well, as shown below.

Edgar Cayce (1877-1943), James Watt, Einstein, Winston Churchill, Mozart, Beethoven, George Washington Carver, Marconi, Henry Ford, Luther, Madame Cure, Edison, Marconi, Socrates, the French patriot

Joan of Arc, 20th century poet Amy Lowell, mystic Hildegard of Bingen, Mohandas K. Ghandi,

Stephen King—The King of Horror

Novelist **Stephen King** gives an account of how dreams help him in his writings in an interview with an English reporter, Stan Nicholis.

Nicholis: If inspiration for *Misery* didn't come from a real-life incident, where did it come from?"

King: **"Like the ideas for some of my other novels, that came to me in a dream.** In fact, it happened when I was on Concord, flying over here, to Brown's. I fell asleep on the plane, and dreamt about a woman who held a writer prisoner and killed him,, skinned him, fed the remains to her pig and bound his novel in human skin. His skin, th writer's skin, I said to myself, 'I have to write this story.' Of course, the plot changed quite a bit in the telling. But I wrote the first forty or fifty pages right on the landing here, between the ground floor and the first floor of the hotel."

"Another time, when I got road-blocked in my novel *It*, I had a dream about leeches inside discarded refrigerators, I immediately woke up and thought, 'That is where this is supposed to go.

"Dreams are just another part of life. To me, it's like seeing something on the street you can use in your fiction. You take it and plug it right in. Writers are scavengers by nature."

Nicholis: "This could explain the line in *Bag of Bones* that goes,
Perhaps in dreams everyone is a novelist."

2.12 A Search for Counter Examples

Can you identify any sum of digits of first n terms of the Fibonacci sequence and consecutive terms of the entire Fibonacci sequence that forms a sequence R_n whose difference of consecutive terms is not equal to a Fibonacci number with the exception of transition points ?

Can you identify any sum of digits of first n terms of the Fibonacci sequence and consecutive terms of the entire Fibonacci sequence that forms a sequence R_n whose difference of consecutive terms is not equal to a Fibonacci number without the exception of transition points ?

Can you identify an even-subscripted Fibonacci number u_k that cannot be expressed as a product of two numbers a Lucas number, a_b and a Fibonacci number u_a

where a = b and k = a + b?

Can you identify an odd-subscripted Fibonacci number u_k that can be expressed as a product of two numbers a Lucas number, a_b and a Fibonacci number u_a

where a = b and k = a + b?

Can you identify a Fibonacci number u_n such that $R_n = u_n + n_k$ where $R_{n+1} - R_n \neq u_{n-1}$?

A search for counter examples here is equivalent to searching for controversies. It is also equivalent to searching for negative intuition. That will be the job of skeptics who may be doubting the authenticity of some of these dreams and similar experiences. Nobody is forcing you to abandon your belief systems for someone else's. Information has been provided, it is now left to you to make informed, intelligent decisions and draw your own conclusions. Let us listen to the skeptics. They are still talking and at the same time searching for counter examples.

Remote viewing, pioneered by scientists in a US government-sponsored program at SRI (Stanford Research Institute) International in Palo Alto, California, allows participants to enter an altered state of consciousness and journey mentally to distant places to gather information. By searching for counter examples, skeptics are saying that the U.S. government, for over 20 years did not use remote viewing to gather intelligence data on everything from hostages to Soviet missiles and bomb-making facilities (*Intuition Magazine, Issue 13, October 1996*).

By searching for counter examples, critics and skeptics are saying that Jean Houston, Penney Peirce, Carolyn Myss, Nancy Rosanoff, Sharon Franquemont etc. are all professional quacks. These are all experts on the subject of intuition whose intuitive abilities have been recognized in many parts of the world.

By searching for counter examples, these critics and skeptics are saying that the so many books out there on the subject of intuition and extraordinary phenomena like near-death experiences, telepathy, clairvoyance, out of body experiences (OBE), etc. should be banned and removed from library and other book shelves. They are also saying that all internet sites with similar information should be pulled down and the owners prosecuted.

By searching for counter examples, skeptics are also saying that Socrates, Joan of Arc, George Washington Carver, Mohandas K. Ghandhi, Martin Luther King, Jr., Winston Churchill, (*The Journal of Transpersonal Psychology*), Mitchell Liester of Colorado Springs) (*With the Tongues of Men & Angels, Arthur Hastings*) reported fake cases of channeling (phenomenon of people hearing and forwarding messages from disembodied spirits—including the Supreme Being (*Intuition Magazine, Issue 18, October 1997, page 27*).

By searching for counter examples, skeptics are also of the opinion that the eureka moment of Archimedes in the bathtub, and Dr. James Watson's work on the molecular structure of the DNA, Dr. Banting's discovery of the basis of insulin while in a dream state etc. should all be dismissed by a simple wave of the hand and not to be believed and recognized as great breakthroughs in science.

Skeptics might be thinking that anyone reporting or having this and similar experiences, is crazy or mentally sick. Rhea White is the founder of Human Experience Network (EHE NEtwork), a non-profit group in New Bern, North Carolina whose primary goal is studying life-changing exceptional experiences.

Let her in her own words say what she thinks about all this: "Exceptional experiences can be shrugged off, suppressed, explained away, or kept secret by the person experiencing them, who may be frightened or afraid she or he will be viewed as strange, weird, or even mentally ill by others." *(Intuition Magazine, Issue 18, October 1997).*

These people fall back to live on the physical realm and have their minds closed to the non-physical. This chapter has a mission and that mission is to encourage them to boldly recognize and cooperate with these experiences, come out of their closets, and be normal again.

Skeptics have quite a lot to say. I can go on and on but let me say this. None of my dreams could be attributable to either alcohol or drug use, or overeating etc. since none is a part of my lifestyle, and I was not on any conventional or traditional medication of any sort at the time that could have interfered with my state of being or sense of reality.

Should the subjects of dream and intuition appeal to you, the following are worth reading:

Suggested Reading

The Mystical Magical Marvelous World of Dreams, Wilda Tanner
The Natural Artistry of Dreams, Jill Mellick, Ph. D
Our Dreaming Mind, Robert L. Van de Castle, Ph. D
All About Dreams, Gayle Delanry
Awakening Intuition, Frances Vaughan
A Layman's Guide to Unlocking the Extra Sensory Power of Your Mind, Ingo Swann

Learn to See: An Approach to Your Inner Voice Through Symbols, Mary Jo McCabe

The Frontiers of the Soul: Exploring Psychic Evolution, Michael Grosso

Awakening the Mystic Gift: The Surprising Truth about What It Means to Be Psychic, Jane Doherty

Exploring the World of Lucid Dreaming, LaBerge & Rheingold

A Thousand and One Nights of Exploring Lucid Dreaming, Lynne Levitan

Lucid Dreaming: A Concise Guide to Awakening in Your Dreams and in Your Life, Stephen LaBerge

Dream Work: Techniques for Discovering the Creative Power in Dreams, Jeremy Taylor

Developing Intuition: Practical Guidance for Daily Life, Shakti Gawain

Becoming Psychic: Spiritual Lessons for Focusing Your Hidden Abilities, Stanley Krippner & Stephen Kieruiff

Other Worlds: Out of Body Experiences and Lucid Dreams, Lynne Levitan & Stephen LaBerge

Dreams for Dummies, Penny Pierce

Dreams Do Come True: Using Your Dreams to Discover Your Full Potential, Layne Dalfen

Healing Sounds From the Malaysian Rain Forest, Roseman Marina

Recurring Dreams: A Journey to Wholeness, Sullivan Kathleen

Dream Language: Self Understanding Through Imagery and Color, Robert J. Hoss

Dream Catcher: A Young Person's Journal for Exploring Dreams, Patricia Garfield

The Mind Race: Understanding and Using Psychic Abilities, Russell Targ and Keith Harary

The Nature of Personal Reality, Jane Roberts

Natural ESP: A Layman's Guide to Unlocking the Extra-Sensory Power of Your Mind

The Intuitive Way

A Guide To Living Your Inner Wisdom, Penny Peirce

Intuitions, Winter

Seeing With The Heart

Your Sixth Sense, Belleruth Naparstek
Unlocking The Power of Your Intuition
Do it Yourself Intuition, Sharan Franquemont
The Intuitive Edge, Philip Goldberg
How to Use Your Gut Instinct for Greater Personal Power, Marcia Emery
The Path to Inner Wisdom, Patricia Einstein
A Guide to Discovering and Using Your Greater Natural Resource
Intuition at Work, Jefffrey Mishlove, Willis Harman, Gary Zukov, Michael Ray, Sharon
Franquemont, Nancy Rosanoff, Roger Franz
Intuition: An Inner Way of Knowing, Doris J. Shallcross & Dorothy A Sisk
Mendeleyev Dream
The Quest for the Elements, Paul Strathern
The Everyday Psychic Book
Tap Into Your Inner Power and Discover You Inherent Abilities, Michael R. Hathaway, D.C.H.
Discover and Develop Your Sixth Sense At Any Age
Psychic Power, Rob MacGregor
The Conscious Universe
The Scientific Truth of Psychic Phenomena, Dr. Dean Radin
The Psychic Workshop A Complete Program for Fulfilling Your Spiritual Potential, Kim Chestney
A Still, Small Voice
A Psychic Guide to Awakening Intuition, Echo Bodine
The Everyday Dream Book, Trish & Bob MacGregor.
The Complete Idiots Guide To Psychic Awareness, 2nd Ed., Lynn A. Robinson, La Vonne Carlson-Finnerty
Be Psychic Now, Nathaniel Friedland
Psychic Abilities: How To Train and Use Them, Marcia Pickards
With This Gift: The Story of Edgar Cayce, Anne E. Neiwark
Get Psychic: Discover Your Hidden Powers, Stacey Wolf
Synchronicity: Signs of Synchronicity, Patricia Rose Upczak
The Grave of Great Things: Creativity and Innovation, Robert Crudin

Chapter 3

My Personal Odyssey in the World of Dreams

3.1 Objectives

At the end of the lesson, the students should be able to:

(a) Realize the importance of keeping dream journals.
(b) Realize that some dreams have purpose and meanings.
(c) Realize that ultimate reality is beyond the five mundane or physical senses.

3.2 Introduction

Listening to the queen of talk shows, Oprah Winfrey, on her show on July 3, 2000 as she discussed the importance of keeping gratitude journal with members of her audience, made me think twice of my dream journal. Some of the guests read and shared their journal entries with the audience. The famous John Edwards (a channeler) has also emphasized the importance of keeping journals as a part of developing one's intuitive abilities. John Edwards has appeared on several national US TV shows including Oprah Winfrey (ABC), Jane Pauley Show (CBS) to mention but a few etc.

3.3 Dreaming & Dream Journaling as Normal Human Experiences

First of all, it is normal to dream dreams and normal also to have a dream journal. In this chapter, I will be sharing some of my dream experiences which include flying without wings. I have flown solo several times and once in a helicopter. I have flown unaided over buildings, crowds of people, natural vegetation etc. The flying dream experience is amazing, at times for consecutive nights. The feeling is comparable to an altered state and it is. All my flying dreams are lucid. In a lucid dream, the dreamer is consciously aware that he or she is dreaming. I have flown once or twice with my sisters, others are solo. Each time I fly the feeling is the same. It is just a kind of feeling you would like to get addicted to and never want to stop. Most of these dreams are nocturnal.

In my dreams, I have also listened to highly sophisticated and subtle musical compositions. Some of my dreams involve numbers, algebraic expressions embedded in both elementary and advanced mathematics. I regard those as extraneous. They are outside our domain of interest. Some of the dreams reported here are dated, while others are not because of poor records due to procrastination. Review your dreams periodically in order to check for correspondence.

3.4 My Dream Journal

Each dream starts with a title (whether the title makes sense or not, it doesn't matter) and the date it occurred. It is better to use a title that best describes the story. The first recorded dream starts with a dream I had way back in 1975. It is titled, "The Asa Road Triangle".

Dream Activities for 1975, 1988, 1993

3.5 The Asa Road Triangle—Friday, January 10, 1975

It was spring of 1975. I was staying with an uncle of mine while preparing for an ordinary level GCE examination. My center then was Ibo National

Secondary School, Aba popularly known as 'Ibonaco' among its students and others from far and wide. In the wee hours of the morning, on Monday, January 13, 1975, I had a dream where I saw on page 2 of January 13, 1975 issue of the then *Renaissance* newspaper (a state-owned newspaper of the then East Central State of Nigeria with capital at Enugu) a notice of entrance examination into Alvan Ikoku College of Education, Owerri, Imo state, Nigeria.

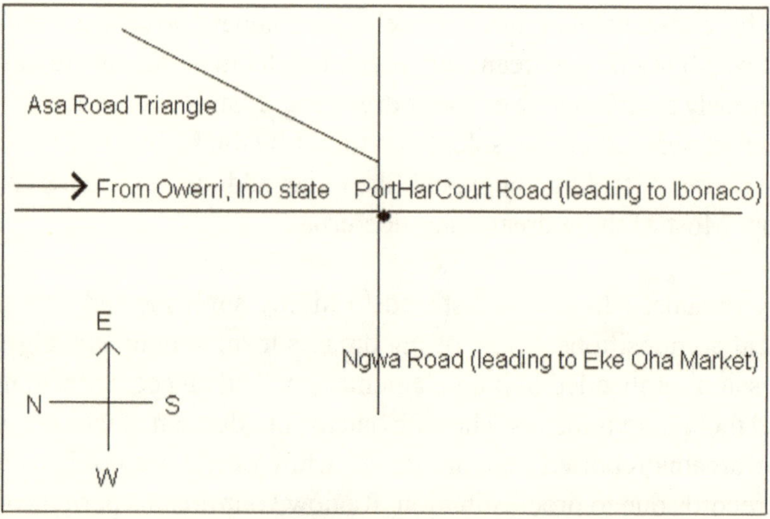

Figure 3.1: Ngwa Road & PortHarcourt Road Junction, Aba, Imo state

The various subject combinations were also displayed in the dream. It should be recalled by some of us that entrance examinations into Nigerian universities, colleges of Education, and other institutions of higher learning were then handled by individual institutions. There was then, neither Nigerian Universities Commission (NUC) nor Joint Admissions & Matriculation Board (JAMB).

Checking for Correspondence

In the morning of January 13, 1975 at the junction of busy PortHarCourt Road and Ngwa Road, I beckoned a newspaper vendor for a copy of the newspaper. To my amazement, on page 2 was the notice of entrance examination into Alvan Ikoku as seen in the precognitive dream three days earlier. My jaw dropped and I stood absolutely dumbfounded. This became my first recorded experience of ESP (Extra-Sensory Perception). I eventually gained admission into Alvan Ikoku two years later.

REMARK: The Asa Road Triangle heightened my interest in the study of the paranormal which includes clairvoyance, clairaudience, out-of-body experiences, mental telepathy, near-death experiences, astral projection, and of course dreams etc. The human mind is very fascinating and just like space, exploring it is an endless endeavor.

Year	Mathematical Dreams *(Frequency)*	Others *(Frequency*
1975	1	0
Total	1	0

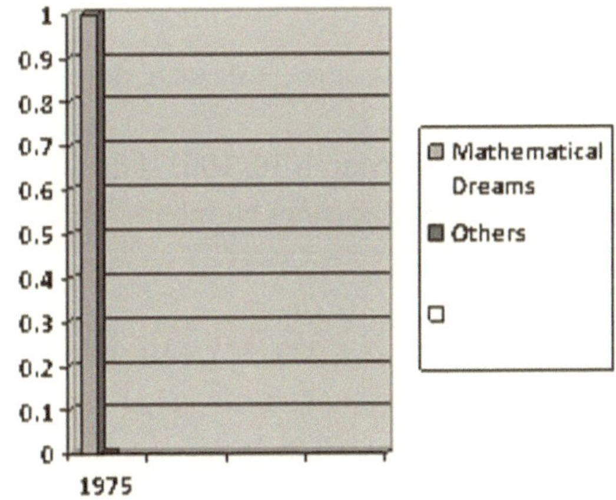

Table 3.1: Summary of Dream Activities for 1975

3.6 Principia Mathematica—Wednesday, January 13, 1988

In this dream, I was sitting for a mathematics examination. On another occasion in the dream I encountered many mathematical formulae that I have never seen in a formal classroom or in a textbook. Unfortunately, none of the information was recollected on awakening.

Year	Mathematical Dreams (Frequency)	Others (Frequency)
1988	1	1
1989	0	0
Total	1	1

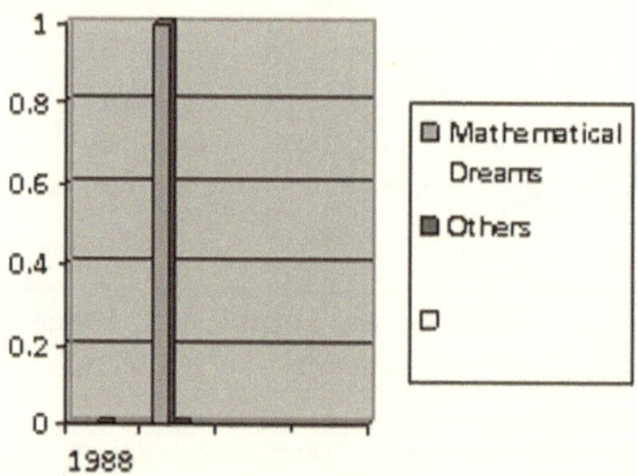

Table 3.2: Summary of Dream Activities for 1988-1989

3.7 The Fascinating Fibonaccis—March 30, 1993

In the wee hours of the night of March 30, 1993, I dreamed about a subset of a number sequence called Fibonacci Numbers. These are numbers of the form:

1, 1, 2, 3, 5, 8, 13, 21, 34, 55, 89, 144, 233, 377, 610, 987, 1597, x, y, x + y . . .

In the dream, they were arranged as white flowers, each flower corresponding to an element (in this case, a Fibonacci number) in the subset.

Checking for Correspondence

Out of this dream experience, and out of sheer inspiration, came a book entirely on Fibonacci Numbers titled, *Fibonacci Numbers for Research Mathematicians & AI Applications*. The book was once sold through the most popular internet bookstore, Amazon.com. Few copies were also sold through Baker & Taylor Books. With a new state of the art technology, the book has been typeset making it look more professional. The title has also been recently changed to *A Course in Fibonacci Numbers.*

What People are Saying about This Book

The book will certainly appeal to teachers and researchers of mathematics. Of great interest is the potential application of the contents of the book to the emerging research efforts in Artificial Intelligence (AI). Fibonacci search is presently one of the search techniques being explored for AI applications. Any book that sheds more light on the unique properties of Fibonacci Numbers should be of interest to AI practitioners and AI researchers. *Dr. Adedeji Badiru, Professor of Industrial Engineering, and Interim Dean, University College, University of Oklahoma.*

Your book was a fascinating exploration into mathematics. *Stacey Weinard, Former Mathematics Coordinator, Oklahoma State Department of Education*

Fibonacci numbers are the numbers of an innocent-looking sequence named after an Italian mathematician, Filius Bonacci. *Fibonacci Numbers for Research Mathematicians & AI Applications* will intrigue students, teachers, researchers in the field of mathematics, as well as those engaged in the research of artificial intelligence and computer science. Author Paul Emekwulu presents a fascinating mathematical exploration that is challenging, illuminating, and original. *Reviewer's Bookwatch, Midwest Book Review*

There are many important aspects of maths learning which are generally neglected in the math education exploration: pattern recognition and formulation of conjectures. This is how Paul Emekwulu presents the subject in this book. Therefore, I recommend to readers: Read this book as a starting point, and you will better understand the more formal approaches. *Opher Liba, Mathematics Educator, Researcher, and Reviewer Jerusalem, Israel*

3.8 Some Applications of Fibonacci Numbers

Fibonacci numbers in mathematics and computer science

Fibonacci numbers are studied in the area of mathematics known as number theory. Fibonaccian search according to Dr. Adedeji Badiru, Head of Industrial Engineering, Tennessee State University, Tennessee is among the search techniques being developed by researchers in the area of computer science. Fibonacci numbers therefore, have applications in the counting of mathematical objects such as sets, permutations, and sequences and to computer science.

Fibonacci numbers and nature

In well-proportioned human beings, the ratio of the length from the navel to the forehead to the length from the navel to the feet is equal to the golden ratio, 1.6180.

The number of petals and sepals of most beautiful flowers are all numbers of the Fibonacci sequence e.g. Alamanda, Hibiscus, Columbine, Black-eyed Susan etc. The arrangement of seeds in the sunflower presents an interesting pattern. They appear to be spiraling outwards both to the left and the right. There is a Fibonacci number of spirals.

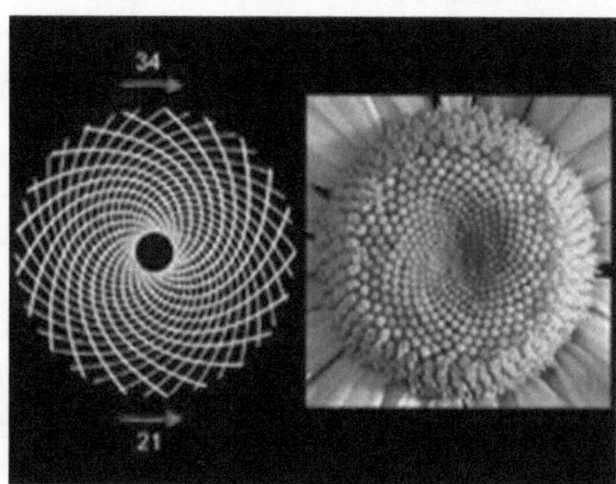

Figure 3.2: Daisy with 21 spirals

Figure 3.3: Columbine with 5 petals

Figure 3.4: Black-Eyed Susan with 13 petals

Fibonacci numbers and Architecture

If we consider two successive terms of the sequence,
$x, y, x{+}y$, then

The golden section is the limit of $\frac{b}{a}$ as n approaches infinity.

So $\text{f} = \frac{1}{\text{f}} + 1$

Clearing fractions we have:

$\text{f}^{2} - \text{f} - 1 = 0$

$$\text{f} = \frac{1+\sqrt{5}}{2} = 1.6180$$

The golden ratio which was known to the Greeks of the old is normally denoted by the Greek letter, φ. The Greek mathematicians of the time of Plato (about 400 BC) recognized it as a significant value and Greek architects used the ratio 1.6180 as an integral part of their designs, the most famous of which is the Parthenon in Acropolis, Athens. The Parthenon was started in 447 BC and finished in 432 BC.

Figure 3.5: The Parthenon in Acropolis

Fibonacci Numbers and the Golden Section

An interesting and a special value closely related to the Fibonacci series, is called the *golden section*. The golden section is obtained by taking the ratio of successive terms of the Fibonacci series. As n gets larger and larger, this ratio approaches an interesting limit. This limit is actually the positive root of a quadratic equation and is called the golden mean, or sometimes the golden ratio or still divine ratio.

The Beauty Equation and Fibonacci Numbers

In a study to measure attractiveness, Dr. Kendra Schmid, assistant professor of Biostatistics in the College of Public Health at the University of Nebraska Medical Center, says there is a "perfect face". Her study used geometry to measure attractiveness on a scale of 1 to 10. She uses the golden or divine ratio and 29 other factors to study what she calls facial sex appeal. Other factors used in the study include symmetry.

The study shows that as the ratio of the length of the face to the width of the face gets closer and closer to the golden ratio, both male and female images were viewed as more attractive.

$$\frac{Length\ of\ \ Face}{Width\ \ of\ Face} \approx 1.6\,.$$

As defined by the golden ratio, the ideal result is 1.6.

Dr. Schmid's study used 36 participants of Caucasian faces (18 males, and 18 females).

Dr. Kendra has been on The Oprah Winfrey Show - Sex: Women Reveal What They Really Want. She is also a part of a two-hour documentary on the Discovery Channel. The documentary is titled, *The Science Sex Appeal.* Dr. Schmid's research was published in Pattern Recognition magazine in August 2008.

3.9 The Healing Fibonaccis—Tuesday, July 6, 1993

I was so fascinated with Fibonacci Numbers that on this night I dreamed of curative or healing Fibonacci Numbers and as defined in the dream are those Fibonacci Numbers that under certain conditions, when uttered generate healing energies in the body capable of providing a cure for certain ailments. Curative Fibonacci numbers are therefore, much like mantras—vowel sounds which when uttered tend to produce some physiological changes in the body as evidenced in transcendental meditation—a technique that centers on the use of mantras and aims to bring about a state of tranquility.

REMARK: To my knowledge, no known research has been done in this area. Any such future research should involve mathematicians, parapsychologists, and any interested stakeholders, and medical doctors with inclination to alternative healing therapies.

Dream Activities for 1994, 1996

3.10 The School Mate Geometry—Tuesday, November 12, 1996

There are two characters in this dream: myself and a school mate of mine in Nigeria who was taking a geometry lesson from me (in the dream).

3.11 Mathematical Solution—Wednesday, November 13, 1996

This is a dream where I was going through a solved mathematical problem.

3.12 The School Mate Algebra—Sunday, November 17, 1996

In this dream, I was exposed to advanced algebra in the presence of a one-time school mate of mine and a friend.

3.13 Mathematical Revelation—Saturday, December 7, 1996

In this dream, I was presented with mathematical questions followed by the solutions.

3.14 The Geometric Model—Saturday, December 14, 1996

In this dream, I was sitting in a class where a female professor was proving a mathematical theorem using a geometric model.

Year	*Mathematical Dreams*	*Others*
1988	*1*	0
1993	2	1
1997	6	0

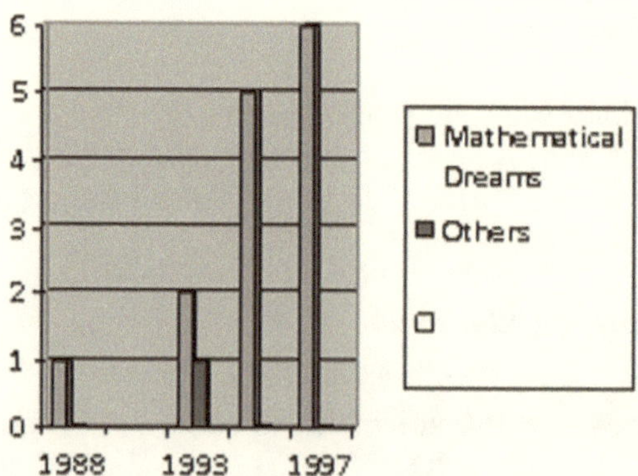

Table 3.3: Summary of Dream Activities for 1988 - 1997

Dream Activities For 1997

3.15 The Sacred Document—Monday, January 6, 1997

In this flying dream, I saw a document with the numbers 45 and 46. Of course, like previous flying dreams, the feeling is still the same.

REMARK: If $45 = n$, then $46 = n+1$.

Sum of n and $n+1 = n+(n+1) = 45+46 = 91$.

91 is the 13th triangular number since $\dfrac{13 \times 14}{2} = 13 \times 7 = 91$.

Multiplying n and $n+1$ we have:

$n(n+1) = 45 \times 46$.

Since n and $n+1$ are consecutive, $\dfrac{n(n+1)}{2}$

should be a triangular number and of course,

$\dfrac{n(n+1)}{2} = \dfrac{45 \times 46}{2} = 1035$.

The number 1035 is the 45th triangular number since $n = 45$. It is interesting to note that within the time I had this dream, I was writing my book on triangular numbers with the title, *A Course in Triangular Numbers*.

3.16 The Publisher—Monday, January 20, 1997

A voice was asking me when I would publish my book on triangular numbers. I made it clear to him that I have other preferences.

3.17 The Mathematics Teacher—Monday, January 20, 1997

Two ladies were asking me whether I would apply to teach mathematics to a group of high school students at a public institution. My answer of course was yes.

3.18 The Transfer Literal Equation—Sunday, March 16, 1997

The characters in this dream were an uncle, Patrick Ifeonu—a high school teacher at Obeledu Girls' High School, Obeledu in Idemili LGA of Anambra State, Nigeria and two boys of high school age dressed in school uniforms. I was asking my uncle whether he was transferred to another school. After the dream I remembered very clearly the following algebra problem that reads exactly as given below in addition to others that I couldn't remember on awakening.

If $s^2 = g^2$, what is st^2, given that $t = g$?

SOLUTION

$st^2 = ?$

If $s^2 = g^2$, then s = g (taking square root of both sides).

By substituting for s in st^2
(since s = g) we have:

$st^2 = gt^2$

$= g.g^2 = g^3$ ($t = g$, given).

Therefore, $st^2 = g^3$ (**).

Verification of the Above Result

To verify our result, we have to work from the RHS back to the LHS in (**) above.

We have to show that $g^3 = st^2$.

$g^3 = t^3$ (since $t = g$, given)

$= t \,.\, t^2$ .... Equation 1

Also $s^2 = g^2$ (given).

Taking the square root of both sides:

$\sqrt{s^2} = \sqrt{g^2} \Longleftrightarrow s = g$

But since t = g and g = s, from here t = s.
By substituting for t in equation 1 we have:

$t.t^2 = s.t^2 = st^2$.

Therefore, $st^2 = g^3$.
This has been verified.

3.19 Sacred Geometry—Friday, April 25, 1997

In this dream, I saw several geometric patterns. These were not recorded because of recall difficulties.

3.20 The Pattern—Saturday, March 29, 1997

In this dream, I saw the two geometric patterns shown in figs. 3.7. I borrowed the title of this dream from a similar incident experienced and reported in an article "Beyond and Back" by Lynnclaire Dennis *(Intuition Magazine, December 1997)* after a near-death experience in 1987.

Figure 3.6: The Pattern according to Lynnclaire Dennis

Both patterns (mine and Dennis') may fall under what is called "sacred geometry" discussed in a separate article titled, "Sacred Shapes" by Victor Beasley, the author of *Intuition by Design* who goes on to say that the use of sacred geometry has existed throughout history in architectural design, various cultures have used the technology of geometry to enhance spiritual consciousness in places of worship such as Egyptian pyramids and Islamic mosques. Sacred geometry, according to Wikipedia (the free encyclopedia) "is the geometry used in planning and construction of religious structures such as churches, temples, mosques, religious monuments, altars, tabernacles, as well as for sacred spaces . . . and the creation of religious art."

"Sacred geometry is an ancient means for creating altered state." Sacred geometry is said to have been used by the Greek philosopher and mathematician, Pythagoras. He had his students focus on three simple geometric patterns—the cross, triangle, and circle—because he had discovered that the act would expand their consciousness. Within the same period of time, I dreamed of more complex and sophisticated geometric patterns that needed highly sophisticated software and technical skills to represent on paper.

Figure 3.7: The Pattern

Figure 3.8: The Basilica of Saint Peter, Rome
(Source: Free Wikipedia)

Dream Activities For 1998

3.21 Le Professeur Mathematique—Monday, January 9, 2006

I thought this was a real examination not until I realized that it was a dream. The subject is again mathematics. The question paper which is light green in color is about two or three pages long. Some minutes or so into the examination, a math professor of average height and who is black, came in and took away question papers from some of the students because of some kind of examination malpractice.

One of the questions is on deductive logic involving Sin 3θ. The questions are many. I don't know the exact number or exactly what branch or branches of mathematics the questions were drawn from. Also, I do not know when I entered the examination hall. I just found myself taking an exam. At some point, I had a need to borrow writing paper from fellow students around me. They did not have the right size of paper. My neighbor on the left started to hand over to me $8\frac{1}{2}\times 11$ sheets. I needed something much more than that in length. Neither did I see the end of the examination nor did I see the beginning but an examination took place.

REMARK: *Le Professeur Mathematique* made one point clear about dream recall. After a dream, stay in bed and get relaxed. At times after a dream you don't remember you had a dream. With some relaxation, brainstorming, visualization, the dream will start to unfold either as a whole or in fragments. This is exactly what happened with *Le Professeur Mathematique*. Recalling a dream this way is always an "eureka moment" for the dreamer.

3.23 The Analog Clock—Thursday, October 1, 1998

It was early morning just a little more than a month after I moved into a new apartment in a flash, I saw a face of an analog clock with the long hand pointing at 12 and the short hand pointing at 6.

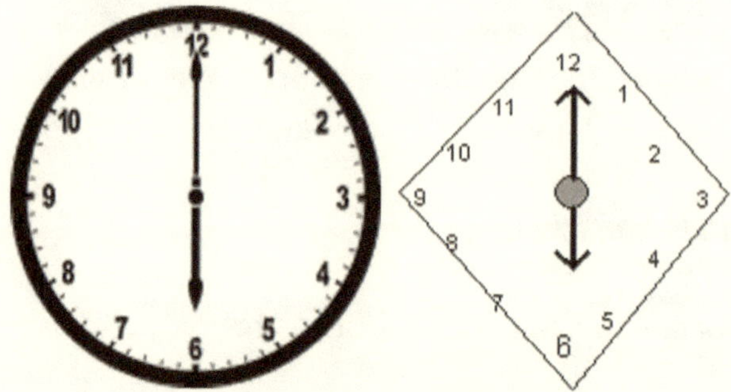

Figure 3.9: An analog clock

Checking for Correspondence

On awakening, the time was 6.00 A.M. (See Figure 3.9).

Dr. James E. Rankin the Chiropractor - 1999

The actual date was not recorded. It was winter of 1999. I have just moved into a new apartment. I formed the habit of sleeping on the sofa and watching late night TV after a hard day's job. In the dream, I saw the phone number, 360-8000. In real life, neither have I seen this number nor have I called it before. I have neither met Dr. Rankin nor do I know him personally. The next morning there was a delayed decision whether to call or not to call the number. "What do I lose?", I told myself. I did not care for an office visit. I eventually discovered on January 17, 2006 that Dr. Rankin's office is located at 828 Wall Street in Norman, on the west side of highway 1-35 South.

Figure 3.10: Location of Dr. Rankin's Office in Norman

Eventually, I decided to call. What I discovered was amazing. A lady answered the call. Her words: "this is chiropractor's office, how can I help you." I was dumbfounded. I told her that I called because I saw the phone number in a dream the night before. She didn't really say anything meaningful in return. Maybe she thought I was crazy and I thought maybe she was another skeptic and she probably was. The call ended. I called 360.8000 to schedule an appointment to visit with Dr. Rankin. I recounted my dream experience to the administrative assistant. She said she would get with Dr. Rankin first, that she cannot just schedule an appointment. I left my phone number with her. She again asked me if I was coming in for a treatment. I said no, it was for my dream experience. She demanded that I tell her what the appointment was for. My call was never returned.

Checking for Correspondence

A second opinion is not really necessary here. Had I continued to sleep on the sofa, I would have had a need to visit a chiropractor's office. This dream is clearly precognitive or predictive. "The number one chiropractic issue in the United States today is back pain," says a doctor with Allied Health Services in Oklahoma City.

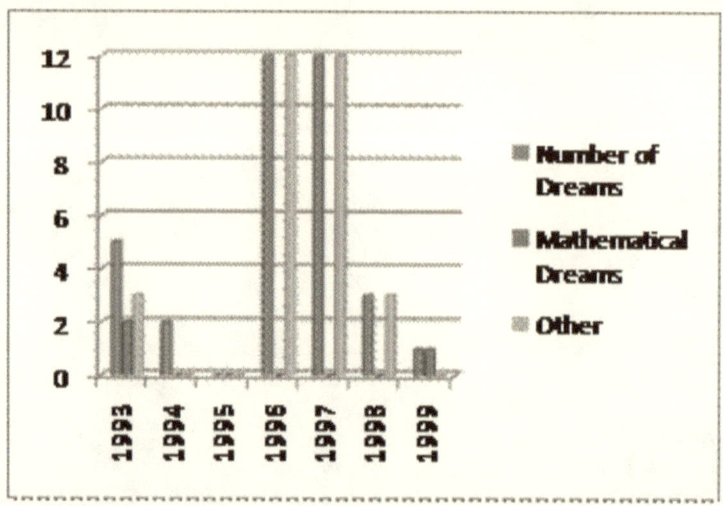

Table 3.4: Summary of Dream Activities for 1993-1999

3.24 The Synchronicity

The date of this dream was not recorded due to procrastination. In this dream, I saw two phone numbers with the following anatomy:

(a) (315)*******   (b) (918) 367-2210

The first area code 315 belongs to the city of Syracuse in the state of New York. The seven stars represent the unknown seven digits which unfortunately, I did not remember. On the other hand, the area code 918 belongs to Tulsa County, one of the 72 counties in the state of Oklahoma. As I checked the Oklahoma State Department of Education School Directory, I found out that (918) 367-2210 used to belong to Caddo Public Schools in Bryan County, Southeastern Oklahoma. As I took a closer look at the area code of 315, and the access code in the second phone number, some interesting patterns started to emerge. Just look at the following:

Sum of First and Third Digits in 315 and 367

In each of the numbers, the sum of the first and third digit is an even number.

$3 + 5 = 8$, $3 + 7 = 10$.

Not only that. These sums are not only even, they are also consecutive.

In 315 and 367, the sum of second and third digits is a square number i.e. $3+1 = 4$, $3 + 6 = 9$, and 4 and 9 are consecutive square numbers.

Product of First and Third Digits in 315 and 367

Again, in 315 and 367, the product of the first and third digit is a triangular number i.e.

$3 \times 5 = 15$, $3 \times 7 = 21$ and 15 and 21 are consecutive triangular numbers.

Sum of Digits in 315 and 367

Also in 315 and 367, the sum of the digits is equal to a square number i.e. $3 + 1 + 5 = 9$, $3 + 6 + 7 = 16$ and 9 and 16 are consecutive square numbers i.e.

$3^2 = 9$, $4^2 = 16$. It should be recalled that the sum of two consecutive triangular numbers is equal to a square number.

Describing the Third Digits in 315 and 367

Each of the numbers has 3 as its third digit while 315 and 367 end with the odd numbers 5 and 7 respectively which are again consecutive. Amazing! Isn't it?

Checking for Correspondence

Since *The Synchronicity,* I have written a book on the subject of triangular numbers. The title is *A Course in Triangular Numbers*. Triangular numbers, it should be recalled are numbers of the form: 1, 3, 6, 10, 15, 21, 28, 36, 45, 55, 66, 78, 91, 105, 120, 136, 153 . . .

The word, 'consecutive' was used 89 times (*not by design*) in *A Course in Triangular Numbers, Book 2*. The number 89 is of course, the 11th Fibonacci number.

What People Are Saying About This Book

In reviewing your book, according to the listed criteria, I have found *A Course in Triangular Numbers* to be mathematically stimulating. Having been a secondary math teacher for 20 years prior to my current position as a secondary education/curriculum specialist, I believe most good math students that successfully complete Algebra 1 could grasp the logical steps in the proofs and patterns. However, maturity is needed to more fully appreciate the significance of the relationships discovered. Therefore, the recommended audience should be AP mathematics students, gifted students participating in an independent study during high school, or post-secondary students. Of course, mathematicians of all levels should be included—teachers, researchers, and theoreticians alike.

I do know of books that deal with triangular numbers as a small chapter, unit, or lesson, mostly in a descriptive context, but I am not aware of any that delve into the patterns and relationships as deeply or thoroughly as your book does. I do confess that I haven't truly researched the subject either. The mathematics is sound and well developed, and the objectives listed for each article greatly support the topic of triangular numbers".

Thank you for the opportunity of reviewing your book. You are an accomplished mathematician and author, and I look forward to working with you in the future.

Penny Jackson, Secondary Education Curriculum Specialist, Lawton Public Schools, Lawton, OK, USA

Dream Activities for 2003, 2004, 2005

3.25 The Amazing Mathematics—Tuesday, June 3, 2003

This is a dream involving advanced mathematics.

3.26 Maxima & Minima—Thursday, January 1, 2004

This is a dream where a mathematics professor was solving maxima and minima problems. He asked me a question and my answers were TeX and Mathematica. These are computer programs for typesetting professional, mathematical and scientific documents. ***Happy New Year***

3.27 The Matrix—Friday, May 21, 2004

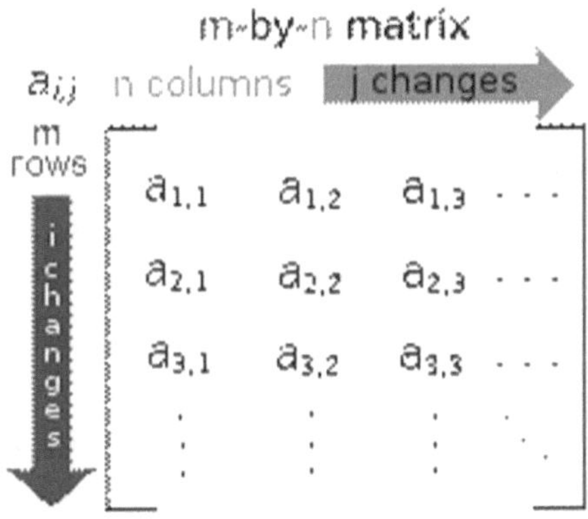

Figure 3.11: General Representation of a Matrix

This is a dream on matrices. Details were not provided

3.28 The Fascinating Algebra—Tuesday, August 9, 2005

In this dream, some algebra problems were unveiled to me. Unfortunately, none of these were remembered upon awakening.

3.29 The Examination—Sunday, October 9, 2005

Here I was sitting for a mathematics examination. On my left is a lady. We all are sitting for this examination. While the examination was going on, she showed me a math book with a yellow cover matching the color of her dress.

3.30 Dr. Iloakagbune, the Geologist—Monday, December 12, 2005

This dream took place at a school. I went there to explore opportunities. As I was standing on the hallway, a good friend, Mr. Hyacinth came to where I was without saying a word he made a sign of the cross and walked away. After a while I saw a man walking enthusiastically towards me from the same direction my friend went to earlier. I recognized him as Dr. Iloakagbune Okeke. We shook hands while the students looked on. We eventually went to a corner where among other things, he asked me whether mathematics was my major course of study.

This was in the presence of two students. Dr. Okeke still looked exactly the way he was back then in the 60s (tall, elegant, and handsome). I promised to show him my various mathematics books which include: *A Course in Fibonacci Numbers*

Mr. Okeke who is a geologist by profession, is a graduate of The University of Nigeria, Nsukka. He said that he heard that I just came back from the United States.

"Yes," I confirmed.

What was not clear in the dream was whether Dr. Okeke was the head of the institution or not.

Checking for Degree of Correspondence

On May 15, 2006, I made a routine phone call to my fiancée in Nigeria. We were casually talking when she asked me how I came to know Dr. Iloakagbune (She read of Dr. Okeke in a review copy of one of my books titled, *Mathematical Encounters For the Inquisitive Mind* that I had earlier sent to her in March of 2006.

"Before I came to the United States," I answered.

"Why do you ask?" I asked her.

"He is my in-law," she said.

"Dr. Iloakagbune is your in-law?"

"Yes," she reinforced.

"I don't believe it." I said to her.

"Yes, he is my in-law. He is married to my senior sister."

My jaw dropped. I was dumbfounded and still in disbelief.

The other character in the dream, Mr. Hyacinth is a good friend of mine who I had earlier called also in December 2005 to help me find a wife.

I am happy to announce that today Patricia Nwabuogochukwu (nee Emenike) is my wife and we have been happily married ever since.

3.31 Ramanujan Algebra—Sunday, December 18, 2005

It was an early sunny Sunday afternoon. I have just came back from work, ate and went to bed to have a short nap. Shortly after, I went to sleep. About an hour or even less into the sleep I had a dream. The setting is a classroom. There was a rectangular piece of paper with a bold inscription on it lying on the floor. A student sitting on my far left and who looks like an Indian (from South-East Asia) thought I was looking at the paper. He reached for it and removed it out of sight.

Now also sitting on my right at the back of this classroom was a class mate I met last in 1980 at Alvan Ikoku College of Education, Owerri. His name is Mr. Samuel Ibezimakọ. He was busy studying while I was rummaging through an old advanced algebra textbook where I saw the most wonderful form of advanced mathematics I have ever been exposed to in my dreaming life. The material presented to me in this dream represents the most fascinating, the most complex, and the most detailed mathematics ever presented to me in a dream state.

Inside the same classroom there was a group of math students listening intensively to a professor who I neither saw nor identified.

A Deep Urge for a Recurring Dream Experience

I would like to have a repeat performance of *Ramanujan Algebra* in the form of a recurring dream, for what was presented to me was very amazing. Relaxed and sitting quietly in front of my personal computer, I reenacted the dream setting to see if I could remember some of the materials but all to no avail.

This dream was named in honor of the Indian mathematician, Srinivasa Ramanujan (1887-1920), who claimed that he received many of his mathematical concepts from the Indian goddess, "Namagiri" etc. Srinivasa Ramanujan was a renounced number theorist.

The thought of *Ramanujan Algebra* sends chills along my spine each time I think about it and I hope it will continue to do so as long as I live. The feeling was a good one.

Dream Activities For 2006

3.32 Continued Division—Saturday, January 21, 2006

In this dream, a classroom environment, a student was working on what looks like continued division. When he failed to get the desired result, I heard myself saying, "for circles, it is fixed e.g. (36, 1), (36, 2), (36, 3) etc." I don't know the reason why I said that and the process involved is not a familiar one.

3.33 The US Citizen— Saturday, January 28, 2006

In this dream I found myself in a room, call it room A full of people enthusiastically, shouting, "It will work", "It will work". Few seconds I found myself in a second room connected to South room by a door. East room is also full of people shouting the same words. Few seconds I picked a piece of chalk and started writing on the chalkboard. The first thing I wrote was the numeral 20 and then the numeral 30 directly below it.

Checking for Correspondence

On Monday, January 30, 2006, the notice of oath Ceremony which was dated January 25 arrived in the mail. The ceremony date is Friday, February 24, 2006. From Wednesday, January 25-Friday, February 24 is an interval of 30 days and this corresponds with the second numeral 30 which I wrote on the chalkboard. This is of course open to argument. It could be a mere coincidence, some might say.

JANUARY 2006

S	M	T	W	T	F	S
1	2	3	4	**5**	**6**	7
8	9	10	11	**12**	**13**	14
15	16	17	18	19	20	21
22	23	24	25	26	27	28
29	30	31				

FEBRUARY 2006

S	M	T	W	T	F	S
			1	2	3	4
5	6	7	8	9	10	11
12	13	14	15	16	17	19
20	21	22	23	**24**	25	26
27	28	29	30			

Date letter postmarked = January 25, 2006
The first numeral written on chalkboard in the dream: 20
25 + 20 = 45
There are 31 days in January. So we subtract 31 from 45.
45 - 31 = 14
The 14th day in February is Valentine's Day.
An oath ceremony cannot be held on a Valentine's Day.
So, 20 days from January 25, 2006 is not a good day for the oath event.
20 doesn't work.
The second numeral written on the chalkboard: 30.
30 + 25 = 55
There are 31 days in January. So we subtract 31 from 55.
Doing so we have: 51 - 31 = 24
55 days from January 25, 2006 corresponds with February 24, 2006.
An oath ceremony can be held on this day since it is not a holiday, nor a special event day like Valentine's Day.

3.34 The Venn Diagram—Friday, February 3, 2006

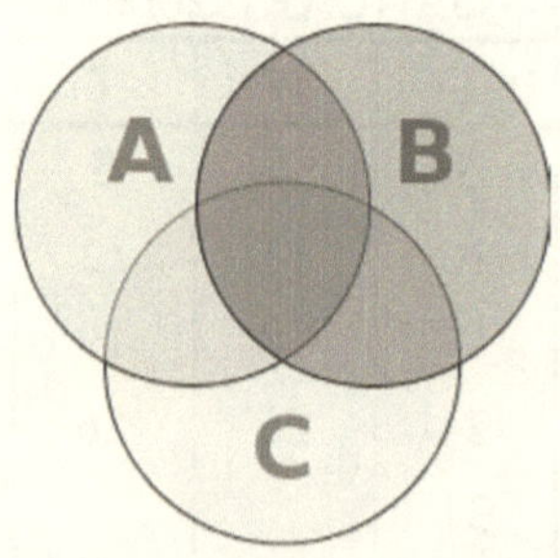

Figure 3.12: A Venn Diagram of sets A, B, C

In this dream (named after John Venn, an English mathematician) I was solving and illustrating problems using Venn diagrams with two unknown classmates, X and Y watching. X called the attention of Y to come and see how such problems are solved using Venn diagrams. Prior to going to sleep I never read any book by John Venn nor engaged in any activities involving Venn diagrams. The concept of Venn diagram was developed by John Venn who lived from 1834-1923. Venn diagrams first appeared in Venn's book, *Symbolic Logic*.

3.35 The Formula Entry—Thursday, March 16, 2006

In this dream a principal of a high school was sharing a list of mathematical topics with a new high school math teacher while I looked on. Among others the topics that were listed was *The Formula Entry*. Others were blurred and less explicit. The topics filled an 8.5 inches by 11 inches sheet of paper and about 20 of them in number. As I looked closer I discovered that the handwriting with which the list was recorded was mine. The only characters in the dream were myself, the new math teacher, and the school principal.

3.36 Father in the Classroom—Monday, March 27, 2006

In the dream, my father was an elementary school teacher. A question came up in the class. I do not remember what the question was but I know it was based on the principle of equivalent fractions for its solution. Only one student got the answer correct. He was able to answer the question because of his ability to apply the principle of equivalent fractions without realizing what he did. I decided to teach the principle to the students. I was very anxious to do this. My father walked away a few yards in the opposite direction.

You can easily tell that my father was tired. Because of this I said, "Father, since you are sick and tired why not get a rest?" I picked up a piece of chalk from the chalkboard. I was about to write when the teacher came back. He regarded me for some seconds without saying a word. I interpreted the silence as meaning consent. Before I could start the lesson, the class had a visitor who dressed in subdued red native attire.

3.37 Dream Sounds Senseless From One Perspective

My father who passed away on April 17, 1995 did not have any elementary or grade school education let alone being taught or having studied the principle of equivalent fractions in school. Seeing him in a dream as a teacher sounds senseless though the dream probably has a message. I never had the opportunity to teach this principle to the students. The characters in the dream include my father, myself, a neighbor whose first name is Louis (years younger than my father), the students, and the class visitor.

3.38 But Makes Sense From Another

The principle of equivalent fractions is a very important principle not only in mathematics but also in other mathematical and biological sciences etc. The principle of equivalent fractions states that if two fractions $\frac{a}{b}$ and $\frac{c}{d}$ are equivalent, then the following are true:

$b \times c = a \times d$	$\frac{c}{a} = \frac{d}{b} \Leftrightarrow ad = bc$
$\frac{a}{c} = \frac{b}{d} \Leftrightarrow ad = bc$	$\frac{c}{d} = \frac{a}{b} \Leftrightarrow ad = bc$
$\frac{d}{c} = \frac{b}{a} \Leftrightarrow ad - bc$	$\frac{a}{c} = \frac{b}{d} \Leftrightarrow ad - bc$
$\frac{a^2bd}{ab}$	$\frac{acb^2}{ab}$
$\sqrt{\frac{ac}{bd}} = \frac{a}{b}$	$\sqrt{\frac{ac}{bd}} = \frac{c}{d}$

Verifying Our Result

Proof that $\sqrt{\frac{ac}{bd}} = \frac{a}{b}$, given that $\frac{a}{b} = \frac{c}{d}$.

Given: $\frac{a}{c} = \frac{b}{d}$

Proof: (i) If $\frac{a}{b} = \frac{c}{d}$, then $c = \frac{ad}{b}$ and $d = \frac{bc}{a}$.

By substituting in $\sqrt{\frac{ac}{bd}}$ for c and d we have:

$$\sqrt{\frac{ac}{bd}} = \sqrt{\frac{\frac{a^2d}{b}}{\frac{b^2c}{a}}} = \sqrt{\frac{a^2d}{b} \times \frac{a}{b^2c}} = \sqrt{\frac{a^2}{b^2} \times \frac{ad}{bc}}.$$

But $\frac{\text{ad}}{\text{bc}} = 1$, since ad = bc.

Therefore, by substituting for $\frac{ad}{bc}$,

$$\sqrt{\frac{a^2}{b^2} \times \frac{ad}{bc}} = \sqrt{\frac{a^2}{b^2} \times 1},$$

so that $\sqrt{\frac{a^2}{b^2}} = \frac{a}{b}$.

Consequently, $\sqrt{\frac{ac}{bd}} = \frac{a}{b}$.

Similarly, we can also prove that $\sqrt{\frac{ac}{bd}} = \frac{c}{d}$

given that the following are true

$$\frac{a}{c} = \frac{b}{d}, \qquad \frac{d}{c} = \frac{b}{a}, \qquad \frac{b}{a} = \frac{d}{c}, \qquad \frac{a}{b} = \frac{c}{d}.$$

3.38 A Search for Counter Examples

Can you identify an equivalent fraction, $\frac{a}{b} = \frac{c}{d}$ such that $a \times d \neq b \times c$?

3.39 The Unknown Tutorial - Thursday, May 4, 2006

In this dream I was organizing a tutorial for about one or two students on an unknown subject with an adult watching. I spoke about having brought out a book on Fibonacci numbers.

3.40 The Proof—Thursday, May 11, 2006

On this day I had two dreams and they were all mathematical. On awakening I was busy trying to remember the first one when suddenly I remembered the second. The second dream was a geometric proof taken from a senior high school geometry syllabus. There were other writings on the board and they were all about geometric proofs.

The setting is a classroom of a higher institution somewhere in Nigeria. The proof outline (let us call it proof outline No. 1) is different from what I was used to when I was in high school in Nigeria (let us call it proof outline No. 2).

Let us take a look at proof outline No. 2 with proofs of the sine and cosine formulae as examples.

Proof of the Sine Formula or the Sine Rule

If a triangle has six angles given the sides can vary in length but still will always be in the same ratio. A triangle has six elements viz:

(a) the 3 sides (b) the 3 angles

It is not known how the ratio of the sides depends on the angles. Will the sides of the 60° angle in a 30°, 60°, 90° triangle be twice as long as the side opposite the 30° angle? The answer is no. To say yes, is suggesting that sides of a triangle are proportional to the opposite sides of a triangle.

What is true? The sides of a triangle are not proportional to the opposite angles but to the sines of the opposite angles and this is what the sine formula is all about.

Let us now take a look at the proof of the sine and cosine formulae.

(A) **Proof of the Sine Formula or Rule**

Given a triangle ABC with AB = c, AC = b and BC = *a*.

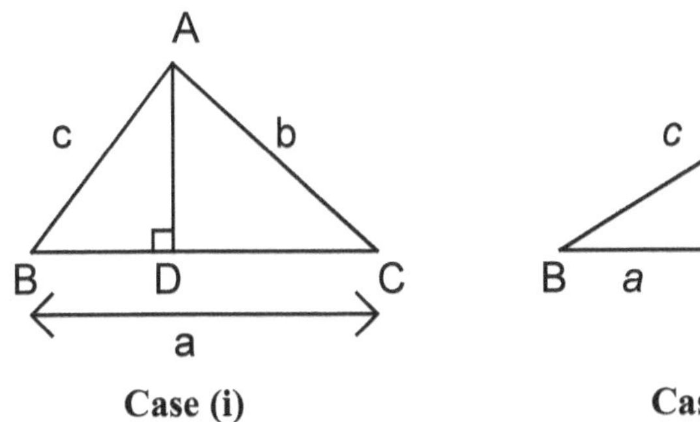

Case (i) **Case (ii)**

(i) $\angle ACB < 90^\circ$
(ii) $\angle ACB > 90^\circ < 180^\circ$

First Case: When the triangle is acute-angled
Construction: Draw AD perpendicular to BC.
Proof: From Δ ABD, AD = c Sin B.
From Δ ADC, AD = b Sin C.
Therefore, c Sin B = b Sin C . . . equation (i)
Dividing both sides in equation (i) by Sin C Sin B we have:

$$\frac{c \operatorname{Sin} B}{\operatorname{Sin} C \operatorname{Sin} B} = \frac{b \operatorname{Sin} C}{\operatorname{Sin} C \operatorname{Sin} B}$$

From here, $\frac{c}{\operatorname{Sin} C} = \frac{b}{\operatorname{Sin} B}$ ....*equation (ii)*

Similarly, by drawing the perpendicular line from B to AC, we can prove that

a Sin C = c Sin A.

Dividing both sides in equation (i) by Sin A Sin C we have:

$$\frac{a \operatorname{Sin} C}{\operatorname{Sin} A \operatorname{Sin} C} = \frac{c \operatorname{Sin} A}{\operatorname{Sin} A \operatorname{Sin} C}$$

From here, $\frac{a}{\operatorname{Sin} A} = \frac{c}{\operatorname{Sin} C}$ ....*equation (iii)*

Since $\frac{c}{\operatorname{Sin} C} = \frac{b}{\operatorname{Sin} B}$ and $\frac{a}{\operatorname{Sin} A} = \frac{c}{\operatorname{Sin} C}$

$$\frac{a}{\operatorname{Sin} A} = \frac{b}{\operatorname{Sin} B}$$

Consequently, $\frac{a}{\operatorname{Sin} A} = \frac{b}{\operatorname{Sin} B} = \frac{c}{\operatorname{Sin} C}$.

Second Case: When the triangle is obtuse-angled

Construction: Draw the perpendicular line from A to meet BC produced at D.

Proof: From Δ ACD, AD = c Sin (180 - C) = b Sin C.
From Δ ABD, AD = c Sin B
Therefore, b Sin C = c Sin B... equation (i)

Dividing both sides in equation (i) by Sin C Sin B we have:

$$\frac{c \operatorname{Sin} B}{\operatorname{Sin} C \operatorname{Sin} B} = \frac{b \operatorname{Sin} C}{\operatorname{Sin} C \operatorname{Sin} B}$$

From here, $\frac{c}{\operatorname{Sin} C} = \frac{b}{\operatorname{Sin} B}$

Similarly, by drawing the perpendicular line from B to AC, we can prove that

a Sin C = c Sin A.

Dividing both sides in equation (i) by Sin A Sin C we have:

$$\frac{a \text{ Sin C}}{\text{Sin A Sin C}} = \frac{c \text{ Sin A}}{\text{Sin A Sin C}}$$

From here, $\frac{a}{\text{Sin A}} = \frac{c}{\text{Sin C}}$

Since $\frac{c}{\text{Sin C}} = \frac{b}{\text{Sin B}}$ and $\frac{a}{\text{Sin A}} = \frac{c}{\text{Sin C}}$,

$\frac{a}{\text{Sin A}} = \frac{b}{\text{Sin B}}$.

Consequently, $\frac{a}{\text{Sin A}} = \frac{b}{\text{Sin B}} = \frac{c}{\text{Sin C}}$

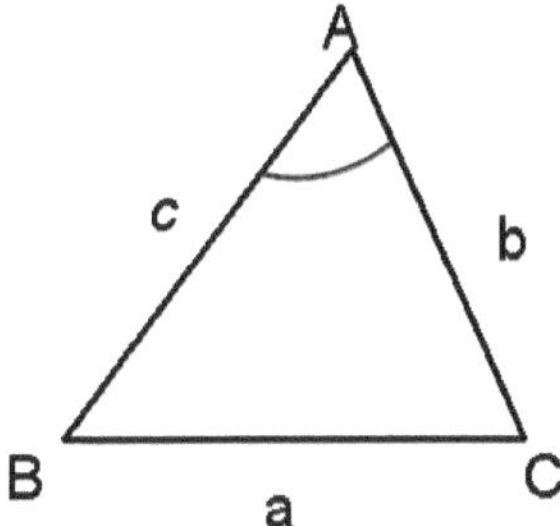

(a) When two sides and an angle opposite one of them are given

(b) When two angles and a side are given

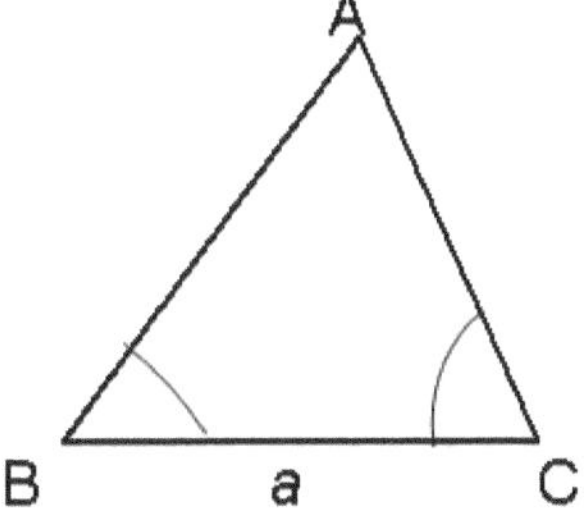

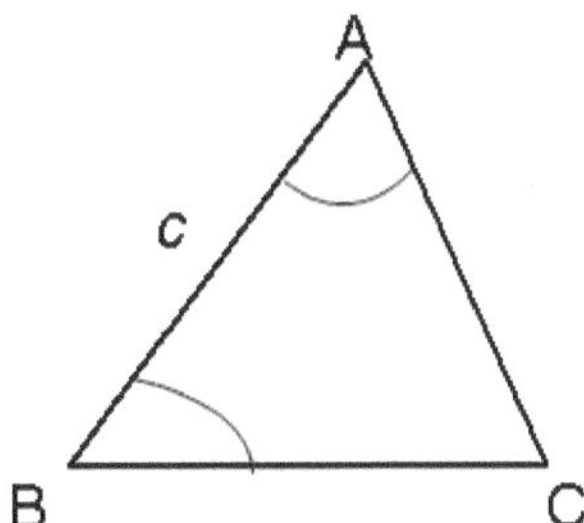

The sine rule can be stated as follows:

In every triangle, the sides are proportional to the sines of the opposite angles.

Conditions for Using the Sine Rule

(a) When two sides and an angle opposite one of them are given
(b) When two angles and a side are given
(i) Never use the sine formula to solve a right-angled triangle. It is not mathematically wrong but is unnecessarily long. (ii) Also do not use the sine formula in solving an isosceles triangle. By drawing a perpendicular from the vertex to the base, an isosceles triangle can be easily solved. Having done that, then proceed to solve the triangle using the Pythagorean theorem. Solving a triangle entails determining an unknown angle or angles or unknown side or sides.

Proof of the Cosine Formula or Rule

Given a triangle ABC with $BC = a$, $AC = b$ and $AB = c$ with:
(i) ACD acute
(ii) ACD obtuse

Construction: From A draw a perpendicular line to meet BC at D as in case (i) or BC produced at D as in case (ii).

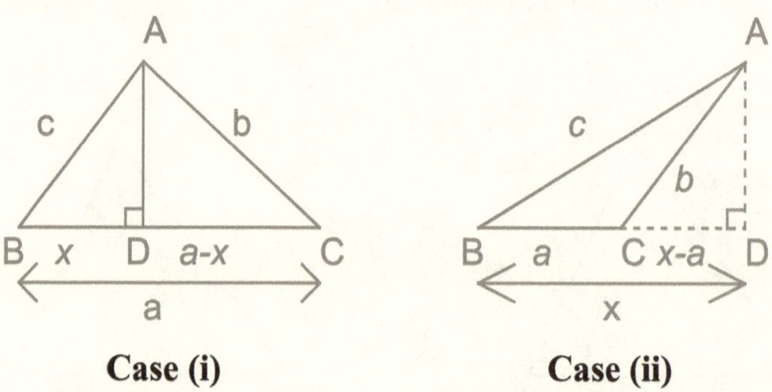

Case (i) **Case (ii)**

Proof: Let $BD = x$
Then CD = a - x, [Case (i), acute-angled]
= x - a [Case (ii), obtuse - angled]

In ⊔ ABD, $AD^2 = AB^2 - BD^2 = c^2 - x^2$ .... Equation 1

In ⊔ ACD, $AD^2 = AC^2 - CD^2$

$= b^2 - (a-x)^2$ (Case i) .... Equation 2

$= b^2 - (x-a)^2$ (Case ii)

Also, $(a-x)^2 = (x-a)^2$

Equating (1) and (2) we have.

$b^2 - (a-x)^2 = c^2 - x^2$

$b^2 - (a^2 - 2ax + x^2) = c^2 - x^2$

$b^2 - (a^2 - 2ax + x^2) = c^2 - x^2$

$b^2 - a^2 + 2ax - x^2 + x^2 = c^2$

$2ax = a^2 + c^2 - b^2$. $x = cCos\,B$

$2ax = 2acCos\,B$ (by substituting for x)

But $2ax = a^2 + c^2 - b^2$

$2acCos\,B = a^2 + c^2 - b^2$ .... Equation 3

Cos B = $\dfrac{a^2 + c^2 - b^2}{2ac}$ **(by dividing both sides of equation 3 by 2ac)**

Similarly, Cos A = $\dfrac{b^2 + c^2 - a^2}{2bc}$

$$\text{Cos } C = \frac{a^2 + b^2 - c^2}{2ab}$$

These formulae can also be expressed in the form:

$$b^2 = a^2 + c^2 - 2ac\,Cos\ B$$

$$c^2 = a^2 + b^2 - 2ab\,Cos\ C$$

$$a^2 = b^2 + c^2 - 2bc\,Cos\ A$$

Conditions for Using the Cosine Rule

When the three sides of a triangle are given and we are required to find any of the angles.

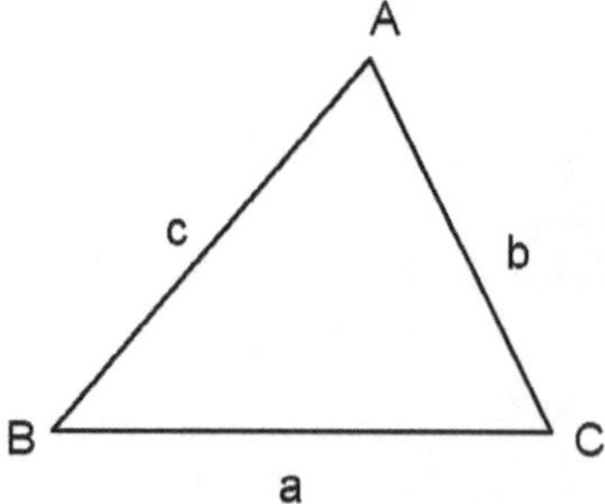

Figure 3.13: When 3 sides are given, required to find any of the angles

When the measure of the included angle and lengths of two sides are given.

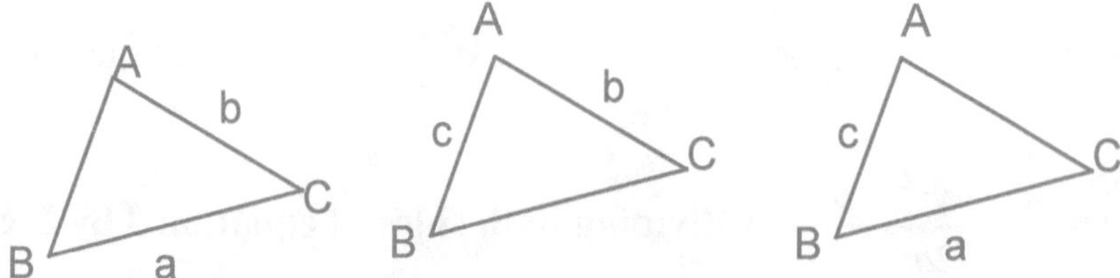

Figure 3.14: When measure of an included angle and two sides are given

3.41 Liber Abachi—Thursday, May 25, 2006

This dream is named after a famous work by Fibonacci, *Liber Abachi* meaning *Book of Calculations* written in 1202 A.D. and was devoted to arithmetic and elementary algebra. In this dream I was sitting on a bench with an unfamiliar male in what looks like a workplace when a lady sitting on the opposite side handed me a hard cover math book 1.5 inches in thickness. The book has an orange cover with the title stamped in gold. Other books by Fibonacci include *Liber Quadratorum and Practica Geometriae.*

3.42 Negative Square Root—Tuesday, August 15, 2006

It was fall of 2006. In this dream in what looks like a mathematics conference, I saw teachers interacting with one another and sharing teaching ideas. After a while I saw a teacher browsing through a math textbook. When I decided to catch a glimpse, I saw a page with the sub topic,

$\sqrt{-pq}$ (square root of a negative number)

where $q < 0$ and $p > 0$ or $q > 0$ and $p < 0$, since if $q < 0$ and $p > 0$, $pq < 0$. (pq is less than zero) or since if $q > 0$ and $p > 0$ since if $q < 0$ and $p < 0$, $pq < 0$)

As we moved away from this part of the conference area, I saw teachers who are in the classroom teaching geometry. In another classroom I saw a chalkboard with the word BOX written boldly on it. Beside the word BOX was a picture of a box.

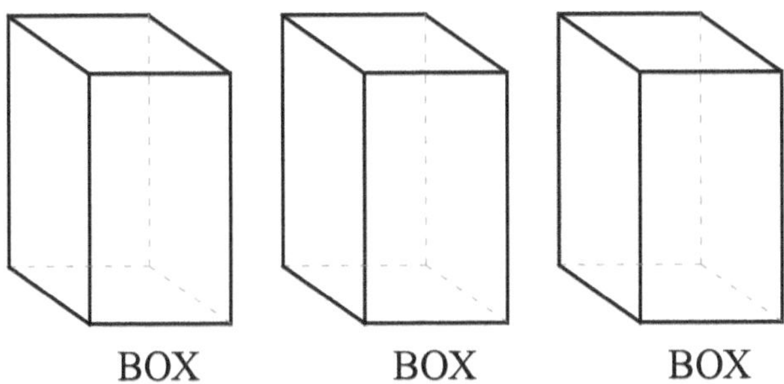

Figure 3.15: Picture of a box

3.43 Graphs of $y = x^3$ and $y = x^2$—Saturday, November 18, 2006

In this dream a boy demonstrated the graph of $y = x^3$ in a public performance while the relatives looked on. After that the boy's father demonstrated the graph of $y = x^2$. When the graph appeared, out of excitement the father demanded an applause from the members of the audience.

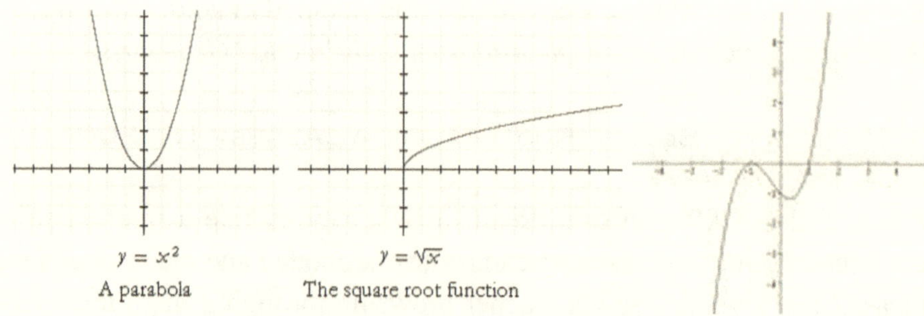

Figure 3.16: Graphs of $y = x^3$ and $y = x^2$

3.44 The Great Mathematics Pamphlet—Saturday, November 18, 2006

In this dream I saw copies of my pamphlet, *Worked Examples & Beautiful Ideas on O' Level Mathematics* lying on a table and some on the floor. I thought they were dropped by students taking their ordinary level GCE examination. Before I could explore further I woke up and it was a dream. This pamphlet sold 100 copies in its first printing and 1000 in its second printing. This was when I was a teacher at Boys' High School, Uga in Anambra State, Nigeria.

3.45 Fibonacci Goes To Las Vegas - Wednesday, December 6, 2006

In this dream I was playing a game with first few terms of the Fibonacci sequence while an unidentified person who asked whether 1000 was a Fibonacci number looks on.

Is 1000 a Fibonacci number? The answer is no.

Why? Because there are no two consecutive Fibonacci numbers, u_n and u_{n-1} such that $u_n + u_{n-1} = 1000$, n ≥ 1.

3.46 The Algebra Class - Wednesday, December 13, 2006

In this dream I was in an algebra class. The teacher had some materials written on the chalk board. He was busy going through this with the class when I called his attention. I noticed an error on what he has written on the board.

3.47 Mathematics Department - Friday, December 15, 2006

In this dream I was at Langston University in Oklahoma asking for direction when a lady met me and asked me what I was looking for and I said, "Mathematics Department." Langston is a predominantly black town north of Oklahoma City and also east of highway I-35 that runs north to south into Canada from US.

Dream Activities For 2007

3.48 The Fraction—Wednesday, January 3, 2007

This dream is about some kind of mathematics involving equivalent fractions e.g.

$$\frac{\mathrm{CCCCn}}{100} = \frac{C}{100} \text{ where } \frac{\mathrm{C}}{\mathrm{C}^4} = \frac{1}{\mathrm{C}^3}$$

3.49 The Proof of the Year - Wednesday, January 3, 2007

If $3ac + \ldots\ldots + 2ac = 7$, prove that . . . The rest of the dream was not clear and before I knew what was happening I woke up.

3.50 Linear Transverse Volume—Monday, January 29, 2007

It was a Monday. I was sitting on a couch watching a TV program. While in a state of half asleep and half awake (hypnotic stage), the words, 'linear transverse volume' flashed through my mind.

3.51 The Meeting of the Minds—Tuesday, February 6, 2007

It was an early Tuesday morning. I was about to have a discussion on the relationship between Fibonacci numbers and triangular numbers with an unknown individual. (Call him Mr. X). Mr. X is a heavily-built man of about 300 lbs and quite at home with Fibonacci and triangular numbers. As I was about to start the lecture, a noise behind me attracted my attention. It was the voice of Mr. X's wife. The discussion never took place and the locale of the dream is not known.

Relationship between Fibonacci and Triangular Numbers

(A) For any three consecutive Fibonacci numbers, a, b, c,

$\frac{b^2+4ac+(c+a)}{2}$ is always a triangular number whose subscript is equal to c + a.

Examples with One Digit Fibonacci Numbers

EXAMPLE 1 Consider the numbers 1, 1, 2.

$a = 1, b = 1, c = 2$

$$\frac{b^2+4ac+(c+a)}{2} = \frac{1^2+(4\times1\times2)+(1+2)}{2}$$

$$= \frac{1+8+3}{2} = \frac{12}{2} = 6$$

EXAMPLE 2 Consider the numbers 1, 2, 3.

$a = 1, b = 2, c = 3$

$$\frac{b^2+4ac+(c+a)}{2} = \frac{2^2+(4\times1\times3)+(1+3)}{2}$$

$$= \frac{4+12+4}{2} = \frac{20}{2} = 10$$

Examples with Two Digit Fibonacci Numbers

EXAMPLE 1 Consider the numbers 13, 21, 34.

$a = 13, b = 21, c = 34$

$$\frac{b^2+4ac+(c+a)}{2} = \frac{21^2+(4\times13\times34)+(13+34)}{2}$$

$$= \frac{441+1768+47}{2}$$

$$= \frac{2256}{2} = 1128$$

EXAMPLE 2 Consider the numbers 21, 34, 55.

$a = 21, b = 34, c = 55$

$$\frac{b^2+4ac+(c+a)}{2} = \frac{34^2+(4\times21\times55)+(55+21)}{2}$$

$$= \frac{1156+4620+76}{2}$$

$$= \frac{5852}{2} = 2926.$$

A Search for Counter Examples

Can you identify any three consecutive Fibonacci numbers a, b, c such that $\frac{b^2+4ac+(c+a)}{2}$ is not always a triangular number whose subscript is not equal to $c+a$?

B) For any three consecutive Fibonacci numbers a, b, c $\frac{b^2+4ac-(c+a)}{2}$ is always a triangular number whose subscript is equal to $c+a$.

Examples with One Digit Fibonacci Numbers

EXAMPLE 1 Consider the numbers 1, 1, 2.
$a = 1, b = 1, c = 2$

$$\frac{b^2+4ac-(c+a)}{2} = \frac{1^2+(4\times1\times2)-(2+1)}{2}$$

$$= \frac{1+8-3}{2}$$

$$= \frac{6}{2} = 3$$

EXAMPLE 2 Consider the numbers 1, 2, 3.
$a = 1, b = 2, c = 3$

$$\frac{b^2+4ac-(c+a)}{2} = \frac{2^2+(4\times1\times3)-(3+1)}{2}$$

$$= \frac{4+12-4}{2}$$

$$= \frac{12}{2} = 6$$

Examples with Two Digit Fibonacci Numbers

EXAMPLE 1 Consider 1 the numbers 3, 21, 34.

$a = 13, b = 21, c = 34$

$$\frac{b^2+4ac-(c+a)}{2} = \frac{21^2+(4\times13\times34)-(34+13)}{2}$$

$$= \frac{441+1768-47}{2}$$

$$= \frac{2256}{2}$$

$$= 1128$$

EXAMPLE 2 Consider the numbers 21, 34, 55.

$a = 21, b = 34, c = 55,$

$$\frac{b^2+4ac-(c+a)}{2} = \frac{34^2+(4\times21\times55)-(55+21)}{2}$$

$$= \frac{1156+4620-76}{2}$$

$$= \frac{5700}{2}$$

$$= 2850$$

A Search for Counter Examples

Can you identify any three consecutive Fibonacci numbers, a, b, and c, such that $\frac{b^2+4ac-(c+a)}{2}$ is not always a triangular number whose subscript is not equal to $c+a$?

Prove that for any three consecutive Fibonacci numbers, a, b, and c, half the product of sum of subscripts of a and c and the sum of subscripts of a and b is equal to a triangular number whose subscript is equal to the sum of subscripts of and b.

Given: a, b, c are three consecutive Fibonacci numbers.

To prove that $\frac{1}{2}(XY) = T_{2n+1}$

where T_n is the nth triangular number.
X = sum of subscripts of a and c.
Y = sum of subscripts of a and b.

Proof

Let a, b, and c be three consecutive Fibonacci numbers with subscripts n, $n+1$, and $n+2$, respectively.

$X = n + (n+2) = 2n+2$
$Y = n + (n+1) = 2n+1$

Therefore, XY = $(2n+2)(2n+1)$.

$$XY = (2n+2)(2n+1) = 4n^2 + 6n + 2$$

Therefore, $\frac{1}{2}(XY) = \frac{1}{2}(4n^2+6n+2) = 2n^2+3n+1$.

Factorizing $2n^2+3n+1$ we have:

$\frac{1}{2}(XY) = (2n+1)(n+1)$ ....(i)

But $2n+1 = n+(n+1)$ and this is the sum of subscripts of a and b.
Let $x = 2n+1$.

Now the xth triangular number is given by:

$$\frac{x(x+1)}{2}$$

By substituting for T_{2n+1} in $\frac{x(x+1)}{2}$, the $(2n+1)^{th}$ triangular number is:

$$\frac{(2n+1)(2n+1)+1}{2} = \frac{(2n+1)(2n+2)}{2}$$

$$= \frac{4n^2+6n+2}{2} = 2n^2+3n+1 = (2n+1)(n+1) \text{ ....(ii)}$$

Since equation (i) and equation (ii) are equal, $\frac{1}{2}(XY) = T_{2n+1}$.

Examples with Two Digit Fibonacci Numbers

EXAMPLE 1 Consider 13, 21, and 34 as three consecutive Fibonacci numbers.

$u_7 = 13$, $u_8 = 21$, $u_9 = 34$,

$a + c = 7 + 9 = 16$
$a + b = 7 + 8 = 15$
$XY = 16 \times 15$

$\frac{1}{2}(XY) = \frac{1}{2}(16 \times 15) = 120$ ....(i)

By substitution, $T_{2n+1} = \frac{(15)(16)}{2} = 15 \times 8 = 120$ ....(ii)

Since equation (i) and equation (ii) are equal, $\frac{1}{2}(XY) = T_{2n+1}$. But 120 is a triangular number whose subscript is 15 and 15 is the sum of subscript of *a* and *b*.

EXAMPLE 2 Consider 34, 55, and 89 as three consecutive Fibonacci numbers.

$u_9 = 34$, $u_{10} = 55$, $u_{11} = 89$,

$a + c = 9 + 11 = 20$
$a + b = 9 + 10 = 19$
$XY = 20 \times 19$

$\frac{1}{2}(XY) = \frac{1}{2}(20 \times 19) = 190$ ....(i)

By substitution, $T_{2n+1} = T_{19} = \frac{(20)(19)}{2} = 10 \times 19 = 190$..............(ii)

Since equation (i) and equation (ii) are equal, $\frac{1}{2}(XY) = T_{2n+1}$.

But 190 is a triangular number whose subscript is 19 and 19 is the sum of subscript of *a* and *b*.

Examples with Three Digit Fibonacci Numbers

EXAMPLE 1 Consider 144, 233, and 377 as three consecutive Fibonacci numbers.

$u_{12} = 144$, $u_{13} = 233$, $u_{14} = 377$

$a + c = 12 + 14 = 26$
$a + b = 12 + 13 = 25$
$XY = 26 \times 25$

$\frac{1}{2}(XY) = \frac{1}{2}(26 \times 25) = 325$ ....(i)

By substitution, $T_{2n+1} = T_{25} = \frac{(26)(25)}{2} = 26 \times 25 = 325$.............(ii)

Since equation (i) and equation (ii) are equal, $\frac{1}{2}(XY) = T_{2n+1}$.

But 325 is a triangular number whose subscript is 25 and 25 is the sum of subscript of *a* and *b*.

EXAMPLE 2 Consider 233, 377, and 610 as three consecutive Fibonacci numbers.

$233 = u_{13}, u_{14} = 377, u_{15} = 610$

a + c = 13 + 15 = 28
a + b = 13 + 14 = 27
XY = 28 ×27

$\frac{1}{2}(XY) = \frac{1}{2}(28 \times 27) = 378$ ....(i)

By substitution, $T_{2n+1} = T_{27} = \frac{(28)(27)}{2} = 14 \times 27 = 378$...............(ii)

Since equation (i) and equation (ii) are equal, $\frac{1}{2}(XY) = T_{2n+1}$.

But 378 is a triangular number whose subscript is 27 and 27 is the sum of subscript of *a* and *b*.

Below is a general proof of the above.
Consider the three consecutive Fibonacci numbers,

u_{n-1}, u_n, u_{n+1}
$X = (n - 1) + (n + 1) = 2n$
$Y = (n - 1) + n - 2n-1$
$XY = 2n(2n - 1)$

$\frac{1}{2}(XY) = \frac{1}{2}[2n(2n-1)] = n(2n-1)$

From a Non-standard Form to a Standard Form

$n(2n - 1)$ is a non-standard form for representing triangular numbers.

We can transform this form to the standard form which is written as

$$\frac{1}{2}(n)(n+1),\ n \geq 1$$

Regardless whether n is odd or even, the nth triangular number is given by

$$\frac{1}{2}(n)(n+1),\ n \geq 1$$

But $n(2n - 1)$ is the general form for representing odd-subscripted triangular numbers if and only if $n \geq 1$.
Let the n in $n(2n - 1) = n'$ so that $n(2n - 1)$ becomes $n'(2n' - 1)$.

Finding a Relationship Between n and n

We can transform $n'(2n' - 1)$ into the general form: $\frac{1}{2}(n)(n+1)$

n	$\frac{1}{2}(n)(n+1)$	n'	$n'(2n'-1)$
1	1	1	1
2	3		
3	6	2	6
4	10		
5	15	3	15

Table 3.5: Finding a Relationship Between n and n′

From Table 3.5, the following are true:

$1+1 = 2,\ 2 \div 2 = 1$
$3+1 = 4,\ 4 \div 2 = 2$
$5+1 = 6,\ 6 \div 2 = 3$

The above division facts can generally be written as:

$$n+1=2n'$$

$$\tfrac{1}{2}(n+1)=n'.$$

By substituting for n' in $n'(2n'-1)$ we have:

$$\tfrac{1}{2}(n+1)\left[2\left(\frac{n+1}{2}\right)-1\right]$$

$$=\left(\frac{1}{2}\right)(n+1)n$$

$$=\frac{n(n+1)}{2}$$

and this is the general form of the nth triangular number.

Proof that for any three consecutive triangular numbers, *a, b,* and *c,* half the product of the subscripts of *a* and *c* and the sum of subscripts of *a* and *b,* is equal to a triangular number whose subscript is equal to the sum of subscripts of *a* and *b.*

...

Given *a, b, c* as three consecutive triangular numbers

To prove that $1/2(XY) = T_{2n+1}$
where T_n is the nth triangular number.

X = sum of subscripts of a and c.
Y = sum of subscripts of a and b.

Proof
Given *a, b, c* as three consecutive triangular numbers.
Let their subscripts be n, n+1, and n+2 respectively.

X = n + (n+2) = 2n+2
Y = n+(n+1) = 2n+1

Therefore, $XY = (2n+2)(2n+1)$.
$XY = (2n+2)(2n+1) = 4n^2 + 6n + 2$

Therefore, $1/2(XY) = 1/2(4n^2+6n+2) = 2n^2+3n+1$ .... Equation 14
Factorizing $2n^2+3n+1$ we have: $2n^2+3n+1 = (2n+1)(n+1)$

Consequently, $1/2(XY) = (2n+1)(n+1)$.

But $2n+1 = n+(n+1) = Y$ and this is the sum of subscripts of a and b.

Finding the $(2n+1)^{th}$ triangular number

(just as for example, $T_3 = [3(3+1)/2] = 6$).

Now the xth triangular number is given by $[x(x+1)]/2$.
Let $2n+1 = x$.

By substituting for $2n+1$ in $[x(x+1)]/2$ we have:
$T_{2n+1} = [(2n+1)(2n+1)+1]/2$.
$[(2n+2)(2n+1)]/2 = (4n^2 + 6n + 2)/2$
$= [2\{2n^2+3n+1)]/2 = 2n^2+3n+1 =(2n+1)(n+1$ .... Equation 15

Since equation (i) and equation (ii) are equal, $1/2(XY) = T_{2n+1}$
Q.E.D.

Examples with One Digit Triangular Numbers

EXAMPLE 1

Consider 1, 3, 6 as three consecutive triangular numbers.
Let T_n be the nth triangular number.

$T_1 = 1,\ T_2 = 3,\ T_3 = 6$

$X = a + c = 1+3 = 4$

$Y = a + b = 1+2 = 3$

$XY = (3)(4)$

$\frac{1}{2}(XY) = \frac{1}{2}(3 \times 4)$

$= 6$ .... Equation 16

By substitution, $T_{2n+1} = T_3 = \frac{(3)(4)}{2} = 6$ .... Equation 17

But 6 is a triangular number whose subscript is 3 and 3 is the sum of subscripts of *a* and *b*. Since equation 16 and equation 17 are equal, $\frac{1}{2}(XY) = T_{2n+1}$

Examples with Two Digit Numbers

EXAMPLE 1

Consider 21, 28, 36 as three consecutive triangular numbers.
Let T_n be the nth triangular number.

$T_6 = 21,\ T_7 = 28,\ T_8 = 36$

$X = a + c = 6+8 = 14$
$Y = a + b = 6+7 = 13$
$XY = (13)(14)$

$\frac{1}{2}(XY) = \frac{1}{2}(13 \times 14) = 91$

By substitution, $T_{2n+1} = T_{13} = \frac{(13)(14)}{2} = 91$ .... Equation 18

But 91 is a triangular number whose subscript is 13 and 13 is the sum of subscripts of *a* and *b*.

EXAMPLE 2

Consider 66, 78, 91 as three consecutive triangular numbers.
Let T_n be the nth triangular number.

$T_{11} = 66,\ T_{12} = 78,\ T_{13} = 91$

$X = a + c = 11+13 = 24$
$Y = a + b = 11+12 = 23$
$XY = (23)(24)$

$\frac{1}{2}(XY) = \frac{1}{2}(23\times24) = 276$

By substitution, $T_{2n+1} = T_{23} = \dfrac{(23)(24)}{2} = 276$................ Equation 19

But 276 is a triangular number whose subscript is 23 and 23 is the sum of subscripts of *a* and *b*.

3.52 Valentine Mathematics—Tuesday, February 13, 2007

It was late Tuesday morning between 10 a.m. and 11 a.m. I was relaxing on a single sofa in my living room and watching one of my favorite TV programs when I dozed off. After a while a friend of mine came up to me with an algebra problem and demanded a solution. It happened that I have solved this problem in the past and the solution is contained in a notebook that contains solutions to other similar problems (in dream state). I then referred him to the notebook and shared the solution with him. Because the dream occurred just a day before Valentine, it is so titled.

3.53 Yes, I am—Tuesday, February 13, 2007

In this dream a man walked up to me and asked if I was the one who wrote a math book. My answer was "yes" and I showed him a copy of *Introducing Mathematics*.

3.54 The Mathematics Professor—Tuesday, February 27, 2007

A professor gave an algebra problem to his students. A student provided a solution on the chalk board. The next thing I saw was the professor's version of the solution on the left hand side of the chalk board. Without giving any reasons, he did not approve the solution provided by the student. I asked him why he did not give any reasons for not approving the student's solution. He still did not give any reasons. The only noticeable difference I noticed in both the student's and professor's solution of the problem was in the use of variables. Where the student used a and b, the professor used m and n. Every other aspect of the solution seems to be basically the same. In the dream the full text of both solutions were very clear on the chalk board.

3.55 Fibonacci Seminars in Nigeria—Saturday, March 3, 2007

I was lying on the couch hoping to just get a rest. After a while I dozed off and the next thing was a dream where someone brought me a letter from a past student of mine. The letter whose entire content was not revealed to me was about Fibonacci numbers. It started like this:

Sir,

....

....

.... number patterns ....

....

Then I was saying to someone telling her that there was ample opportunity for someone like me as far as Fibonacci numbers are concerned. The opportunities exist in form of organizing seminars on Fibonacci numbers for the Nigerian academic community back home in Nigeria because perhaps some members of this community do not know much about this beautiful, and innocent-looking sequence. The same person asked what it will take for one to do that and I said anyone who has taken some college math courses and a deep interest in number theory in general and of course an equal interest in Fibonacci numbers in particular. The letter though has a beginning, had no official ending.

3.56 Vantage Algebra—Sunday, March 18, 2007

The setting of this dream is a classroom. In the dream a student gave some algebra problems to the teacher who was never identified in the dream. The teacher wrote down these problems sequentially. By then the blackboard was not visible from the corner where I was. As soon as the teacher finished writing I positioned myself at a vantage position from where I was then able to see the teacher's writing. Then someone among the students asked whether the problems were mathematically correct, and my answer was yes.

The same day a friend of mine called me from California. He said, "a co-worker gave me an algebra problem and said it was a part of her son's homework and I thought I would be able to provide a solution." (but he couldn't) He read the problem to me over the phone and requested a solution. Here is the problem and next is the solution.

Solve for x, y, and z in the following equations:

$$3x+2y+z=42$$
$$2y+z+12=3x$$
$$x-3y=0$$

MY SOLUTION

This is a set of simultaneous equations in three unknowns x, y, and z and could be solved by using the substitution method. We can then use the same substitution method to verify the solution.

$3x+2y+z=42$ ....(i)
$2y+z+12=3x$ ....(ii)
$x-3y=0$ ....(iii)

From equation (iii) $x=3y$.
Call this equation (iv).
By substituting for x in equation (i) we have:

$3(3y)+2y+z=42$.
$9y+2y+z=42 \Longleftrightarrow 11y+z=42$ ....(v)

Call this equation (v).
Also by substitution for x in equation (ii) we have:

$2y+z+12=3(3y)$

$2y+z+12=9y \Longleftrightarrow 7y-z=12$ ....(vi)

Collecting equations (v) and (vi) we have:

$11y+z=42$
$7y-z=12$

...
$18y=54 \Longleftrightarrow y=3$.

By substituting in equation (iii) for y we have:
x - 3(3) = 0
From here, $x = 9$.

Now by substitution for y in $7y-z=12$, we can find z.

Doing so we have: $7(3)-z=12$. i.e. 21-z = 12.
From here, z = 21 - 12 = 9.
Answers: $x = 9, y = 3, z = 9$

Another Perspective

$$3x+2y+z=42$$
$$2y+z+12=3x$$
$$x-3y=0$$

Rewriting the above equations we have:

$3x+2y+z=42$ ....(i)
$-3x+2y+z=-12$ ....(ii)
$x-3y=0$..(iii)

Subtracting equation (ii) from equation (i) we have:

$3x+2y+z=42$ ....(i)
$-3x+2y+z=-12$..(ii)
. . .
$6x+0+0=54$.

From here, x = 9.

By substituting for x in equation (iii), $9-3y=0 \Leftarrow 3y=9 \Leftarrow y=3$.

By substituting for x and y in equation (i) or (ii) we can find z.
$3(9)+2(3)+z=42 \Longleftrightarrow 33+z=42$.

From here, $z=9$.

Another Perspective

In an effort to solve this problem using elimination method, I discovered that the equations could further be solved as follows:

$3x+2y+z=42$ ....(i)
$2y+z+12=3x$ ....(ii)
$x-3y=0$ ....(iii)

Rewriting the above equations we have:

$3x+2y+z=42$ ....(i)
$0x+2y+z=3x-12$ ....(i)
$x=3y+0z=0$ ....(i)

From equations (i) and (ii), $2y+z=3x-12$
and from equation (i), $2y+z=42-3x$.

3x - 12 = 42 - 3x
$6x=54 \Longleftrightarrow x=9$.

By substituting in equation (ii), we can find y.
Doing so, in equation (iii), $9=3y \Longleftrightarrow y=3$.
From equation (i) 3x+2y + z = 42.

$(3\times 9)+(2\times 3)+z=42$
$27+6+z=42$
$z=42-27-6=9$

Therefore, x = 9, y = 3, and z = 9.

Yet Another Perspective

Also $2y+z+12=3x$
$42-2y-z=3x$
3x - 12 = 42 - 3x
$4y+2z=42-12=30$
$4y+2z=30$
$2y+z=15$ ....(ii)

Also, $(2y+z)+12=3x$

By substitution for $2y+z$ we have:
$(2y+z)+12=15+12=27=3x$ since $2y+z=15$.

From here, x = 9.

If $x-3y=0$, (see equation) then by substitution,
$9-3y=0 \Longleftrightarrow y=3$

By substitution in (ii), $6+z=15$.

From here, $z=9$.
Answers: x = 3, y = 9, z = 9

3.57 The Pythagorean Rule—Wednesday, March 21, 2007

The setting is a classroom. In the dream, a man said that someone in the office (not known) wants anyone to come to the office and prove the Pythagorean theorem. I said, "let's go." The Pythagorean theorem states that in a right-angled triangle, the square on the hypotenuse is equal to the sum of the squares on the other two sides.

Figure 3.17: Bust of Pythagoras (570 BC-495 BC)

$BC^2 = AC^2 + AB^2$ (see Figure 3.17).

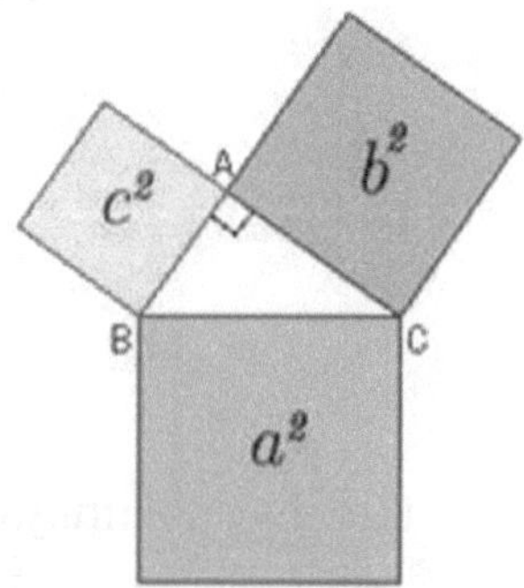

Figure 3.18: Demonstrating the Pythagorean Theorem

Proof of the Pythagorean Theorem

Given a triangle ABC right-angled (Figure 3.19) at A with $BC = a$, $AC = b$, and $AB = c$

To prove that $a^2 = b^2 + c^2$

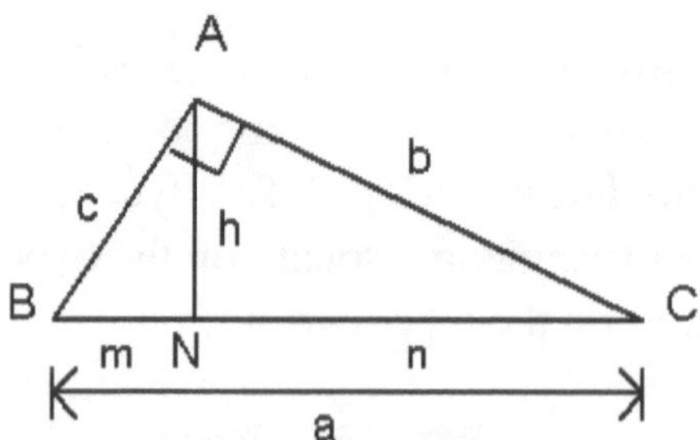

Figure 3.19: ABC right-angled at A with BC = a, AC = b, AB = c, ABC = 60°, ACB =30°

Construction: From A draw an altitude AN to meet BC at the point N. Let $BN = m$ and $CN = n$.

Proof: With the notation in Figure 3.19, the following are true:

In ΔABN, $\text{Cos B} = \frac{m}{c}$ ....(i)

and also in ΔACN, $\text{Cos C} = \frac{n}{b}$ ....(ii)

Also in ΔABC, Cos B $= \frac{c}{a}$ .... (iii)

$\frac{m}{c} = \frac{c}{a} \Longleftrightarrow c^2 = am$ .... (iv)

and in ΔABC, Cos C $= \frac{b}{a}$ .... (v)

Also $\frac{n}{b} = \frac{b}{a} \Longleftrightarrow b^2 = an$ .... (vi)

By adding equations (v) and (vi) we have:

$$am + an = c^2 + b^2$$
$$am + an = a(m + n) = c^2 + b^2$$
$$a^2 = c^2 + b^2 \ (m + n = a).$$ QED.

3.58 Presidential Mathematics—Saturday, April 14, 2007

It was a classroom setting. The teacher was President George W. Bush Jr. He was busy doing some mathematics on the blackboard. The board was not well cleaned because of a previous writing. I had the urge to walk to the blackboard and clean it, but never did and did not know why. Before anything else happened, I realized that I was just dreaming. In the dream I was not the only one present. There were unfamiliar faces and there was neither teacher-student nor student-teacher interaction. The President was saying something to himself. This is my second dream involving the 43rd President of the United States.

3.59 First Annual Fibonacci Lecture—Thursday, April 26, 2007

In the dream someone from an unknown employer was going through my two-paged resume. He really took a lot of time going through this document. After a while he called a co-worker to take the resume and give it more close attention. Few minutes after that I found myself among an adult audience giving a lecture on Fibonacci numbers. I started the lecture by stating the first 19 Fibonacci numbers as follows:

1, 1, 2, 3, 5, 8, 13, 21, 34, 55, 89, 144, 233, 377, 610, 987, 1597, 2584, 4181.

Before then I explained that 3, 6, 9, 12, 15, 18, 21, 24, 27, etc. is a simple number series. In between the lecture, someone came in with some kind of announcement thereby interrupting the class. After that I asked, "how many of you would like me to continue with my lecture on Fibonacci numbers?" They all started laughing. Before the dream ended the next thing I had in mind to discuss with the audience was applications of Fibonacci numbers. Of course that never happened. This was the time I realized that I was dreaming.

3.60 Introducing Mathematics—Friday, May 4, 2007

In the dream a friend who I showed a copy of one of my books, *Introducing Mathematics* was telling me about the book's relevance. I made him to understand that the book was prepared with a sophisticated computer software. He, also a student user of the book talked very highly of it (the software). In another development, I told him that I will never give up fighting on behalf of Math-Magic.

Math-Magic is a program that uses creative and innovative teaching strategies to make mathematics exciting, interesting, and intriguing to high school students using paper and pencil activities. Most of these activities are embedded in guided discovery lessons that come in worksheet format.

The dream concluded with the following:

In mathematics, I see elegance.
In mathematics, I see harmony.
In mathematics, I see consistency.
In mathematics, I also see inconsistencies.

3.61 The Road Side Mathematics—Monday, July 23, 2007

It was not a regular classroom. The characters in the dream were myself, a math professor whose name was not mentioned in the dream, and a student named Chris. Chris was reading out a list of math assignment problems on advanced algebra while the professor was busy providing solutions on the blackboard. At the end of one of the solutions, ∃ was repeated twice and ∃ is a symbol for "there exists."

3.62 The Temperature Scales—Thursday, July 26, 2007

In the dream someone was talking about his preference for Celsius scale of temperature compared to the Fahrenheit scale. He also said something but I couldn't remember what it was. In the dream I asked the following question:

At what temperature are the Centigrade and Fahrenheit scales of temperature equal?

I also mentioned the conversion formulae—formulae for converting Celsius degrees to Fahrenheit and vice visa.

The conversion formulae are as follows:

$C = \frac{5}{9}(F - 32)$ .... (i)

From equation (i) we can find F in terms of C. Doing so we have:
Let the temperature be equal to x.
At x° the following is true:

$C = \frac{5}{9}(F - 32) = \frac{9}{5} + 32 = \frac{9c+160}{5}$

But $F = C$.
Therefore, $\frac{5}{9}(F - 32) = \frac{9}{5}F + 32$

From equation (ii) $\frac{5}{9}(C - 32) = \frac{9}{5} + 32$

$$\frac{5C - 160}{9} = \frac{9C + 160}{5}$$

By cross multiplication we have:

$25C - 800 = 81C + 1440$

$81C - 25C = -800 - 1440$

$56C =^{-} 2240$

$$C = \frac{-2640}{56} = -40$$

Therefore at x = -40°, the two scales of temperature are equal.

Another Perspective

Let the temperature be equal to x.
At x the following is true:

From equation (i), $\frac{5}{9}(F-32) = \frac{9}{5}(F+32) = \frac{9c+160}{5}$

Therefore, $\frac{5}{9}(F-32) = \frac{9}{5}F + 32$

From equation (ii) $\frac{5}{9}(C-32) = \frac{9}{5}(C+32)$

$$\frac{5F-160}{9} = \frac{9F+160}{5}$$

By cross multiplication we have:
$25F - 800 = 81F + 1440$
$81F - 25F = -800 - 1440$
$56F =^{-} 2640$
$F = \frac{-2240}{56}$ = -40°.

Therefore at x = -40°, the two scales of temperature are equal.

3.63 The Black Algebra Textbook—Saturday, August 4, 2007

In the dream I bought and paid for an advanced algebra textbook. The book whose price was not specific was paid for in US dollars. The book has a black thick cover, about $10\frac{1}{2}$ inches $\times 8$ inches in size and also about one and a half inches thick.

3.64 Deductive Proof—Wednesday, August 8, 2007

It was mid-morning on Tuesday of August 8, 2007. This is a dream involving 1/e and using deductive logic to prove that a certain algebraic quantity with the unknown *n* in its denominator is equal to 1/e.

$$\frac{Numerator}{Deno\min ator \times n} = \frac{1}{e} \Leftrightarrow Numerator \times e = Deno\min ator \times n$$

3.65 September Mathematics—Monday, September 3, 2007

I am happy and excited because for some time now I have not had any mathematical dreams. So this is the first of such dream for the month of September. In this dream the subject of examination is mathematics. After this examination as I was looking at other students' answer sheets, I discovered to my amazement that I was given a different examination. As my class mates and I looked closely, we also discovered that my examination was more difficult than the rest of my mates. The only familiar faces in the dream are two high school mates of mine. I wanted to go and report this case to the supervisors but I was busy looking for my answer sheet. Before I realized what was happening, I came to find out that I just dreamed.

3.66 Relative Speed—Saturday, November 3, 2007

It was in the wee hours of the morning at my house in Nigeria. In the dream a friend of mine, a practicing attorney came with a question on "relative speed". I tried to explain the word 'relative' speed with 'reluctant force' in Physics. Before I could demonstrate that, I realized that I was dreaming. Also, I remember mentioning the concept of "resolution of force about a point". At the time I had this dream I was in my room in the United States of America. *Relative Speed* is my first dream in November since my last in September.

Problems on Relative Speed

Problem 1: A car travels at a uniform speed of 45 mph and another travels in the opposite direction at a uniform speed of 55 mph. How far apart are they in 3 hours?

SOLUTION: Relative speed = (55 + 45) mph = 100 mph.

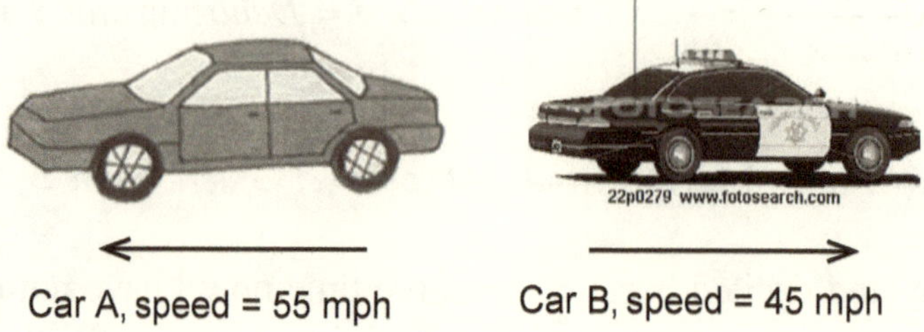

In 3 hours, the distance between the cars = 3 × 100 = 300 miles.

Therefore, in 3 hours, the cars are 300 miles apart.

N.B. Relative speed in opposite direction is the sum of individual speeds, while relative speed in the same direction is the difference of individual speeds.

3.67 Base Notation—Saturday, November 3, 2007

This dream occurred at about 11:30 p.m. In the dream I saw a set of mathematical questions and the first was on base notation.

Dream Activities for 2008

3.68 The Mathematical Novel—Friday, April 4, 2008

The exact time this dream occurred is not known. But it was in the wee hours of the morning of April 4, 2008. In the dream I saw an English novel with half of a specific page devoted to some algebra. This is the first mathematical dream I have had since coming back from Nigeria on February 27, 2008.

3.69 The Questions—Thursday, September, 2008

In the dream I saw a list of questions on mathematics more than two pages long and typed on light green paper. Each question is a word problem. One of the questions was on ratio.

3.70 The Independent—Wednesday, October 1, 2008

It was between 4 a.m. and 5 a.m. on Wednesday, October 1, 2008. I was asking a head student of a high school why the bell had not rang in a long time. I forgot what his reply was. As I asked this question, he allowed me to browse through his notebook. On opening this notebook what I saw was amazing. I saw solved mathematical problems. As I turned the pages backwards I saw more solved problems. I could not remember anything.

3.71 The Young Professor—Friday, October 3, 2008

In the dream a young mathematician probably in his late 20s wrote quite a lot of mathematical writings on the blackboard while two math professors and 1 looked on with interest. When he finished, I volunteered to wipe the blackboard. "I have not done this (referring to wiping the blackboard) in a long time," I said.

3.72 The Pre-inauguration—Friday, January 16, 2009

It was a classroom setting. The lesson was on Higher Algebra that needs using deductive logic to arrive at its proof. After the class, three female students, who by accent are Europeans approached me and asked for my math lecture note. I told them that my note was not complete; that I still have to put the pieces together.

Deductive logic is a method of reasoning developed by Aristotle where a conclusion follows necessarily from certain premises. For example, given the two statements:

All polygons have three or more sides (statement no. 1).
A triangle has three sides (statement no. 2).

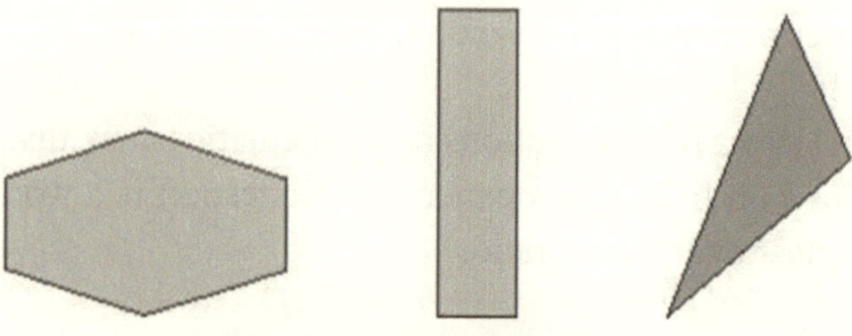

Figure 3.20: Polygons

In deductive logic, it necessarily follows that:

A triangle is a polygon

Deductive reasoning moves from general to particulars while inductive logic moves from particulars to general. Since this dream was on January 16 four days before the inauguration of the 44th President of the United States of America (President Barack Obama) it was so titled.

3.73 Literal Equation—Friday, February 6, 2009

A voice in the dream said that she was not taught literal equations while in high school. "Do you mean making x the subject of the formula?" "Yes," she said. An example of a literal equation was given in the dream but I forgot it on awakening.

3.74 The Unfamiliar Mathematics—Sunday, March 22, 2009

It was a cold Sunday night. In the dream I was browsing through a higher algebra textbook. The content was very exciting and I tried hard when I woke up to remember the details of the dream but all was to no avail.

3.75 The Square Root—Tuesday, April 14, 2009

In this dream, quite unlike in *Unfamiliar Mathematics,* the content of the dream was shown to me very clearly. It was a list of problems in some form of higher mathematics.

3.76 Fourth Grade Measurement—Monday, September 27, 2010

In the dream were myself, my fourth grade teacher, whose first name is Mr. Anthony (now retired), and a third person unknown to either of us. In the dream my teacher was using a cutting machine to trim a slab of thick cardboard paper (see below). He has made several attempts without success (doing all that with a smile on his face). He wanted the cut to be exact. With a smile I said to him, "Every measurement is an approximation. You cannot cut it exactly. That was what I was taught in school."

He continued to smile without saying a word. Everything turned out to be a dream. I woke up at about 3:10 a.m. and wrote the dream down.

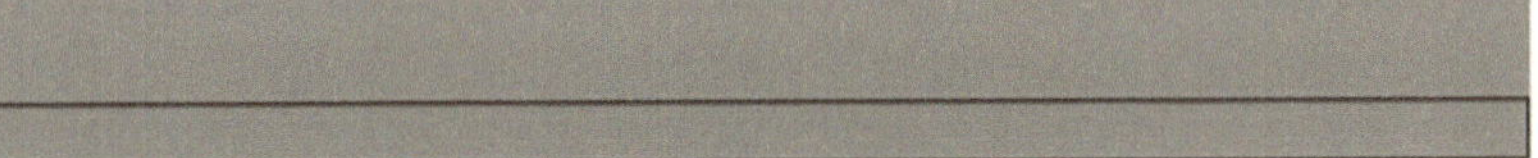

Figure 3.21: Cardboard Paper Slab

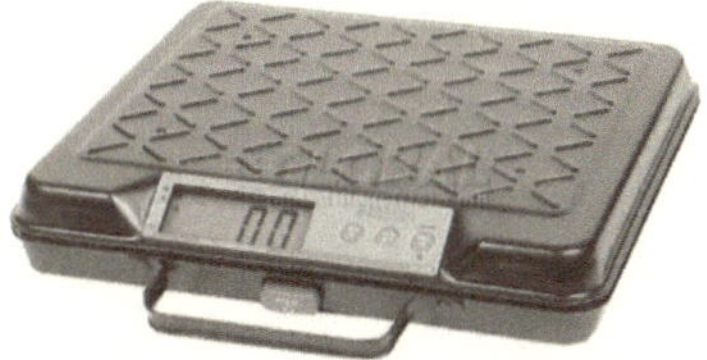

Figure 3.22: Scale

Figure 3.23: Metric ruler

Figure 3.24: Temperature Scale

Figure 3.25: Analog Clock Face

3.77 The Missing Answer Sheet - Friday, October 8, 2010

The setting of this dream is a classroom. I have just finished taking a mathematics examination. When it was time to submit my papers, I discovered to my surprise that my paper was missing. I looked for it almost every where possible all to no avail. I was convinced that it was stolen by another student. I later went back to sleep,. Few minutes into my sleep it recurred. I never found my missing answer sheet. Everything was a dream.

Year	Mathematical Dreams *(Frequency)*	Other *(Frequency)*
2003	1	3
2004	2	0
2005	4	20
2006	14	20
2007	20	0
2008	4	4
2009	4	0
2010	2	0
Total	**51**	**47**

Table 3.6: Summary of Dream Activities for 2000 – 2010

Personal Strategies for Dream Recall

It is easier for me to recall a dream if I wake up as soon as the dream terminates. At times all you need in order to remember a dream is a symbol, a word, a name, an imagery (could be an animate object e.g. a human being, or an inanimate object e.g. a car, or a sentence etc.)

Once any of these makes a connection to the dream, it becomes an 'aha' moment: "Oh, that's it." Here is an example from personal experience. On the 16th of January, 2006, I had a dream. I did not get up as soon as I woke up; I relaxed. I had forgotten everything about the dream and started working on the computer over 12 o' clock news. It was not until I heard the newsman mention the word, 'woman' that I remembered the dream I had few minutes or hours earlier. A second example is a dream I had on 'missing wallet'. I forgot I had this dream not until I saw my wallet lying on the table.

Write the dream down immediately and as detailed as you can remember. Don't worry about grammar and so on. You can easily come back to that. **(2)** When you can't remember a dream, continue to ask questions, engage in brainstorming your mind.

Do not repeat my mistakes. If you decide to keep a dream journal, please be serious about it. No procrastination. Important dates and other details could be lost.

Dreaming Positively About the Future

In an article titled, "Intuitive Dreaming", Alan Vaughan, author of *The Power of Positive Prophecy* gives us the following guidelines for dreaming positively about the future:

Retiring at night, tell yourself that you'll have a positive dream about your future.

(2) On awakening, lie in bed a few minutes and recollect the dream. You may probably recollect the last part of the dream first. Keeping asking yourself: Now what came before that? **(3)** Write down the dream with a title. **(4)** Tell a friend your dream without referring to your notes. **(5)** Look for references to time or to the future in your dream. They may indicate that the dream is about the future. If your dream refers to only to a past event, you know that it is not about the future. If the dream is about some situation that has never happened, it's possible it's about the future. **(6)** Once you have grasped the meaning of the dream, ask yourself: Is there anything I should do now to make this positive future come true? **(7)** Start a dream journal. Periodically review your dreams to compare with recent events and note any correspondences.

Renowned neuroscientist Candace J. Pert, author of *Molecules of Emotion: Why You Feel the Way You Feel.*" has this to tell us about dream recalling and transcription.

Get into a daily habit of recalling and transcribing your dreams because they are direct messages from your body/mind giving you valuable information about what's going on physiologically, as well as emotionally.

A Search for Counter Examples

Can you identify a triangle XYZ right-angled at Y with sides XY = z, YZ = x, and XZ = y such that $y^2 \neq z^2 + x^2$?

By searching for counter examples, skeptics are saying that any information not gained through the conventional four walls of a classroom should not only be questioned but rejected outright and thrown out through the window and the seeker in search of truth prosecuted and imprisoned for life. By searching for counter examples, a skeptic is saying:

I believe in the doctrine of manifest reality.
If I cannot see it, I don't believe it;
If I cannot hear it, I don't want to know about it;
If I cannot taste it, don't tell me about it either;
If I cannot touch it, it doesn't exist;
If I cannot smell it, don't even talk about it and just leave me alone.

I feel comfortable with my present level of existence and don't bother me with multi-sensory perception or multidimensional perspective of reality and all that.

By searching for counter examples, skeptics are still saying that all tapes, videos, books, programs, etc. on the subject of intuition, dreams, and similar phenomena should be dumped into the Atlantic or just burnt to ashes, and all experimental or research laboratories for these phenomena should be burnt or closed down and boarded up.

Can you identify a triangle ABC with AB = c, BC = b, AC = b such that:

$$\frac{a}{Sin\ A} \neq \frac{b}{Sin\ B} \neq \frac{c}{Sin\ C}?$$

Can you identify a triangle ABC with AB = c, BC = b, AC = b such that:

$$\frac{Sin\ A}{a} \neq \frac{Sin\ B}{b} \neq \frac{Sin\ C}{c}?$$

Or

$$a\ Sin\ B \neq b\ Sin\ A \neq c\ Sin\ C \neq b\ Sin\ C \neq c\ Sin\ B \neq a\ Sin\ C \neq c\ Sin\ A?$$

Chapter 4

Fascinating Geometry of the Tangram

4.1 Objectives

At the end of the lesson, the students should be able to:

a) construct the seven piece tangram puzzle
b) name and describe the various shapes of the seven piece tangram
c) identify the relationship between the various shapes of the seven piece tangram

4.2 Introduction

We will take a look at how to construct the seven pieces of the tangram. Next we will provide a description of the seven pieces. After this, we will present the puzzle. The seven piece tangram is an old Chinese manipulative puzzle and consists of seven geometrical pieces constructed originally from a square ABCD.

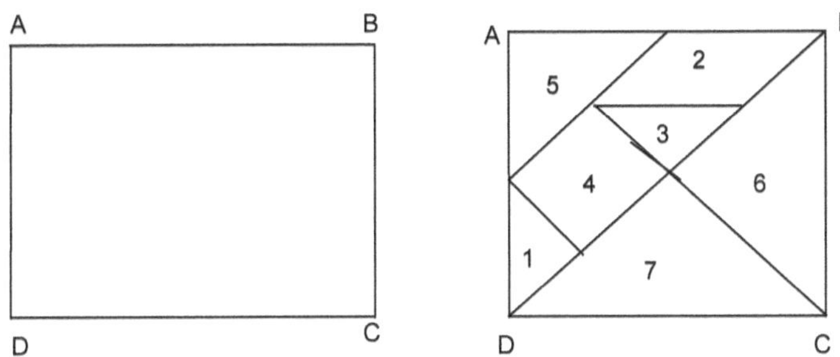

Figure 4.1 Constructing a seven-piece tangram

4.3 How to Construct the Seven-Piece Tangram

a) Construct a square ABCD with each side measuring *a* units.
b) Join the diagonal BD.
c) Find the midpoint of sides AB, AD, and diagonal BD.
d) Let these midpoints be I, E, G respectively.
e) Find the midpoints of DC, BC, and let these midpoints be F and H respectively.
f) Join IE and EF. (g) From the point C, draw a line CJ meeting BD, IE at C and J respectively.
g) Join JH. Name the shapes from 1 through 7 as in Figure 12.1.

4.4 Describing the Seven Pieces of the Tangram

The seven pieces of the tangram could be described as follows:

- Two large congruent right-angled isosceles triangles
- Two small right-angled, congruent, and isosceles triangles
- A medium sized right-angled and isosceles triangles
- A small square
- A parallelogram

4.5 Where is the Puzzle?

The seven pieces are cut along the lines of construction. The pieces are re-shuffled and the objective is to put them back together to form the original square, ABCD.

4.6 Relationship between the Various Shapes

Personally, what I admire about the tangram is the elegant demonstration of some geometric relationships and concepts. I hope you will find them interesting and exciting as I did. First, let us investigate some of these relationships.

4.7 Finding the Unknown Sides

To successfully continue with our investigation, we have to find the unknown sides. In Figure 4.1, AB = CD = AD = BC = *a* units of length. Now by using the notation in the figure, to find the unknown sides, we have:

$$(IE)^2 = \left(\frac{1}{2}a\right)^2 + \left(\frac{1}{2}a\right)^2 = \frac{2a^2}{4}$$

Therefore, $IE = \dfrac{a\sqrt{2}}{2}$.

Since EJ is half of IE, EJ $= \frac{1}{2}\left(\dfrac{a\sqrt{2}}{2}\right) = \dfrac{a\sqrt{2}}{4}$
= IJ = GH = DF = FG

$BD = a\sqrt{2}$

Add the areas of the seven pieces.

What do you notice?

But area of original square is *a* x *a* square units of area.

What does this imply?

4.8 Observations and Conclusions

There are many such relationships between the various shapes of the tangram. Again, from these relationships, many geometric relationships can be found with respect to congruency, similarity, and areas. These relationships therefore, provide some educational insights for the classroom teacher.

From the investigation, we can state as follows:

a) That all similar figures are not necessarily congruent.
b) Shapes 1 and 5 are similar but are they congruent?

The answer is no.

c) All congruent figures are similar.
d) That all similar figures of equal regions (areas) are not necessarily congruent.
e) What of the converse of (c)—congruent figures are equal in area?

Is this true? Which of the shapes among the seven pieces demonstrate the above properties (a - d)?

4.9 A Lesson on Similarity

If two triangles ΔABC and ΔEFC are similar, then the ratios of their corresponding sides are equal.

Considering ΔABC and ΔEFC and with the notation in Figure 4.2,

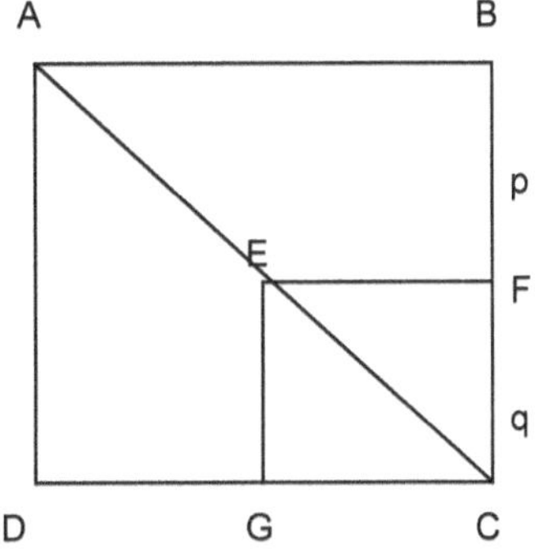

Figure 4.2: A Lesson on Similarity

$$\frac{AB}{EF} = \frac{BC}{FC} = \frac{AC}{EC} = \frac{p+q}{q}$$

The tangram beautifully illustrates one of the theorems involving similarity of triangles which states that if two triangles ΔABC and ΔEFC are similar then the ratio of their areas is equal to the square of the ratio of their corresponding sides.

4.10 Illustrating With Examples

First, let us consider ΔABC in the big square (see Figure 4.2).

To find AC

With the notation in Figure 1, and using the Pythagorean theorem,

$AC^2 = AB^2 + BC^2$

$= (p+q)^2 + (p+q)^2[AB = AC,\text{ side of square}]$

$= (p^2 + 2pq + q^2) + (p^2 + 2pq + q^2)$

$= 2p^2 + 4pq + 2q^2 = 2(p^2 + 2pq + q^2)$

Therefore, $AC = \sqrt{2(p^2 + 2pq + q^2)}$

$= \sqrt{2}\sqrt{(p^2 + 2pq + q^2)}$

$= (p+q)\sqrt{2}$

To Find EC

Now for ΔEFC (see Figure 4.3)

$$(EC)^2 = (EF)^2 + (FC)^2$$

$$= q^2 + q^2 = 2q^2.$$

From here, EC = $q\sqrt{2}$.

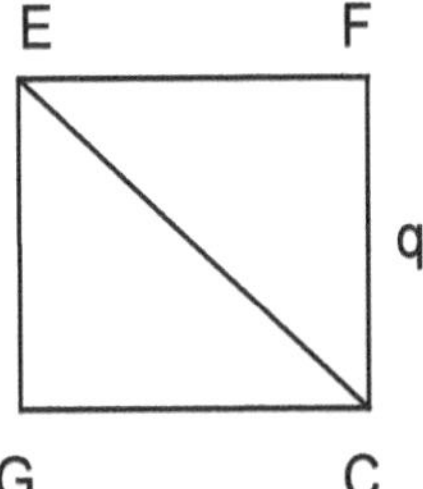

Figure 4.3: Finding EC

Ratios of areas of ΔABC and ΔEFC

$$\frac{\frac{1}{2}(q\times q)}{\frac{1}{2}(p+q)(p+q)} = \left(\frac{q}{p+q}\right)^2$$

Ratios of squares of corresponding sides

$$\left(\frac{FC}{BC}\right)^2 = \left(\frac{q}{p+q}\right)^2 = \frac{q^2}{(p+q)^2}$$ and

$$\left(\frac{EC}{AC}\right)^2 = \left(\frac{\left(q\sqrt{2}\right)^2}{2(p+q)^2}\right) = \frac{2q^2}{2(p+q)^2} = \frac{q^2}{(p+q)^2}$$

$$\left(\frac{FC}{BC}\right)^2 = \left(\frac{EC}{AC}\right)^2 = \left(\frac{AB}{EF}\right)^2 = \left(\frac{p+q}{q}\right)^2$$

Now if we are to substitute for p and q where

$$q = \frac{a\sqrt{2}}{4}, \quad p + q = a$$

and using the appropriate notation, then we have:

$$\frac{Area\ of\ \Delta EFC}{Area\ of\ \Delta ABC} = \frac{\frac{1}{2}\left(\frac{a\sqrt{2}}{4} \quad \frac{a\sqrt{2}}{4}\right)}{\frac{1}{2}a^2}$$

$$= \frac{a^2}{16} \times \frac{2}{a^2} = \frac{1}{8} \text{ .... Equation 1}$$

Now again, $\left(\frac{EC}{AC}\right)^2 = \frac{q^2}{(p+q)^2}$,

By substitution, $\left(\frac{EC}{AC}\right)^2 = \frac{\left(\frac{a\sqrt{2}}{4}\right)^2}{a^2}$

$$= \frac{2a^2}{16} \times \frac{1}{a^2} = \frac{1}{8} \text{ .... Equation 2}$$

Equations 1 and 2 are equal.

Therefore, the two ratios are equal.
If Δ^s ABC, EFC are similar, the ratio of their areas is equal to the squares of the ratio of their corresponding sides. Now again,

$$(EC)^2 = \sqrt{\left(\frac{a\sqrt{2}}{4}\right)^2 + \left(\frac{a\sqrt{2}}{4}\right)^2}$$

$$= \sqrt{\frac{2a^2}{16} + \frac{2a^2}{16}} = \left(\sqrt{\frac{4a^2}{16}}\right) = \frac{a}{2}$$

$$(EC)^2 = \frac{\left(\frac{a}{2}\right)^2}{\left(a\sqrt{2}\right)^2}$$

$$= \frac{a^2}{4} \times \frac{1}{2a^2} = \frac{1}{8}$$

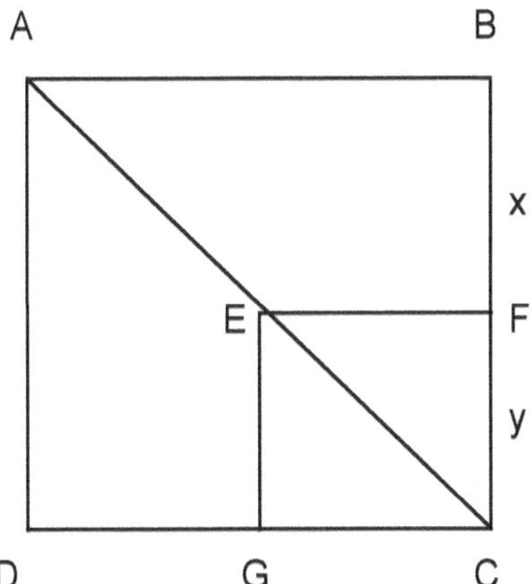

Figure 4.4: Rectangles EFCG, ABCD

The area of the small square in figure 4.4 is one third the total area. Calculate the ratio of $y{:}x$.

SOLUTION

Given $\frac{Area\ of\ rect.EFCG}{Area\ of\ rect.ABCD} = \frac{1}{3}$

For the square of the ratio = $\left(\frac{y}{x+y}\right)^2$

$9y^2 = (x+y)^2$

Taking the square root of both sides we have:

$3y = x + y \Leftrightarrow 3y - y = x$

$2y = x \Leftrightarrow \frac{2y}{x} = 1$

Dividing both sides by 2:

$\frac{y}{x} = \frac{1}{2}.$

Required ratio = $\frac{1}{2}$.

So, $y : x = \frac{1}{2}$.

4.11 A Search for Counter Examples

a) Can you identify any figures that are congruent but not similar?
b) Can you identify any figures that are congruent but not equal in area?
c) Can you identify any figures that are not necessarily congruent but equal in area?
d) Can you identify any figures that are not necessarily congruent but are similar?
e) Can you identify any two similar triangles, ABC, EFC whose ratio of areas is not equal to the square of the ratio of their corresponding sides?

Chapter 5

From Murder Scene to Building and Transforming Word Problems into Simple Equations

5.1 Objectives

At the end of the lesson, the students should be able to:

(a) Improve their mathematical and logical thinking skills.
(b) Use variables in transforming word problems into simple equations.

5.2 Introduction

"I am retiring from modeling today," announced the 31-year-old super model, Tyra Banks on her TV talk show, "Tyra Show". Tyra's career which started at age 16 came to an end on November 21, 2005. She extolled both older and younger black models and then passed on the torch to future and also younger hopeful black super models.

"I am 31 years old. I have been modeling half my life. I am retiring. You are the future. You have to carry the torch," she said to Celina, the youngest black super model. She also gave thanks to some black super models before her who she said paved the way for her. They include Cindy

Crawford, Beverly Johnson. Tyra started her modeling career at a time when the industry seemed to be a preserve of white women and a taboo for black women and other minorities in America.

She shared with her enthusiastic audience her final walk at the 2005 Victoria Secrets Fashion Show. Tyra and her mum exchanged roles. While Tyra's mum became the host, Tyra picked up the role of a guest. She was interviewed by her mum for a fleeting moment. Her mum had arranged a surprise visit by Tyra's close relations, hair dresser and her make-up artist. They all had a good time.

"It is time to go before they kick me out," Tyra said.

5.3 Too Sensitive an Issue

Also, actor Sean Austin is 32. "Oprah Winfrey is 40 today," a radio talk show announced. Oprah Winfrey is one of the richest women in the communication and entertainment industries and was in 2003 declared the first African-American female billionaire. Her TV show known as "The Oprah Winfrey Show" airs throughout the United States and about 64 foreign countries.

"I will be fifty years old tomorrow," an Oklahoman council woman, Willie Johnson said in an interview on a local Oklahoma radio program.

I said all that to say this. To most new African immigrants in the United States talking about one's age is a sensitive issue. To these new immigrants, celebrating birthdays is therefore, not a popular culture. For most Americans, the attitude is different. Americans are more liberal and therefore, at the ready to do otherwise. There could be cultural and social reasons for this reluctance on the part of Africans. The children of these new immigrants who were born in the United States will find it less sensitive if at all to do otherwise. Now, can we move on?

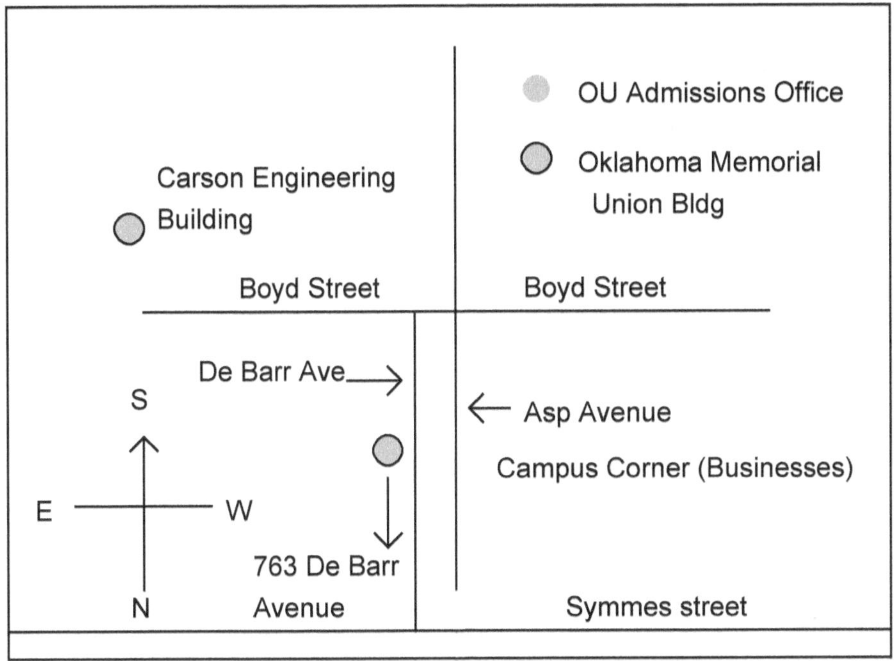

Figure 5.1: Area Map

5.4 Turning Down an Interview at a Murder Scene

It was evening and the year was 1996. A murder has just occurred at 763 De Barr Avenue, Norman where I was living at the time. A man's body has been decapitated. The incident attracted news editors from local print media and a state newspaper, *The Daily Oklahoman & Times*. The editor from *The Daily Oklahoman & Times* wanted to interview me since I was a resident at the apartment. I turned down the request. Why did I turn down the interview? Because I have better things to be interviewed on than a murder case. Few years before then, I had an exceptional psychic experience with a number sequence known as Fibonacci sequence. The same newspaper came and interviewed me. It was an hour and a half long interview. The text of the interview never saw the light of the day (It was never printed). My experience resulted in a book titled, *A Course in Fibonacci Numbers.*

After I left the newsman, I found myself a few yards from Boyd Street opposite Carson Engineering Building near the north oval of the University of Oklahoma (OU) on the same side of Campus Corner on Asp Avenue to welcome a Nigerian friend who came to check on me because of the incident. A few minutes after that, a man, let's call him Mr. Smith probably in his forties came to join us. You can easily tell from his accent that he is not a foreigner. Mr. Smith is American-born who thinks that my friend and I have an accent and not only that we have an accent but a thick Nigerian accent.

5.5 Recognized and Non-Recognized Accents

It is all about what I will call recognized and non-recognized accents (these are mine). Recognized accents are tolerable, while non-recognized accents are intolerable. Recognized accents are found among people from most Western European countries like Britain, Germany, and France. While non-recognized accents are found among people from African countries like Nigeria, Ghana, Liberia, Tanzania, Angola; and Asian countries such as India, China, Pakistan, and Malaysia. Though for some of these countries like India, Malaysia, and China, some people are of the opinion that accent doesn't necessarily matter anymore because of their emerging economies and relevance in world affairs. People from Mexico, Central and South America also share non-recognized accents as well.

Some Americans think that when you have an accent, (a) you don't know the American language, and (b) you have nothing to offer. That is not necessarily so and that has indeed been proven wrong. Professionals from some of these countries are found throughout the United States making a tremendous difference in hospitals, universities, research laboratories etc. Some are even under-employed while some others are engaged in private businesses. Having an accent, as I always say is a matter of location. An American in the busy streets of Surulere, Lagos; Dugbe market, Ibadan; Ekwulobia, Anambra state (all in Nigeria), Mpraeso, Ghana; Freetown, Sierra Leone; Kuala Lumpur, Malaysia; Nairobi, Kenya to name a few surely has an accent.

5.6 Queen's English or Victorian Accent

Some Americans are of the opinion that Nigerian accent is closer to British or Victorian accent. That still doesn't make any difference, for you still have an accent (and remember, a thick one).

5.7 Accent is a Matter of Geography Not Choice

In the United States, as long as you are a foreign-born, you have an accent, your years of US residency notwithstanding. Even the celebrity body builder, multimillionaire, movie star, five-time winner of Mr. Universe title, former Gov. Arnold Alois Schwarznegger (1947-), a naturalized US citizen and an Austrian immigrant, who struggled his way up to become the governor of California is not excluded. Governor Arnold is not the first foreign-born governor. John Downey (1827-1894) was California's first foreign-born but seventh governor from 1860-1862. He was born in Roscommon County, Republic of Ireland. The Southern California city of Downey is named after him. Arnold's political and everyday social speeches are punctuated with his native Austrian tongue. Even if you are a naturalized US citizen, you are still considered as having an accent? Oh, yes, even if you are a naturalized citizen, and also even if you are fastly losing your native accent, the mental attitude is still the same. (Being a naturalized US citizen has nothing to do with your accent anyway.)

In contiguous United States (excluding and Alaska and Hawaii), those living in Oklahoma for example, consider those living in New York (New Yorkers) as having an east coast accent while New Yorkers consider Oklahomans as having an 'okie' (pronounced 'oky') accent. Accent is not a universal. It has a limit and that limit is determined by location. It is not a matter of choice. It is simply a matter of geography.

5.8 That's Simple Arithmetic

"How long have you been in the United States?" (Question no. 1) Mr. Smith asked my friend few minutes into the conversation.
"Twelve years," he answered.

After a few minutes he asked him again.
"How long were you in Nigeria?" (Question no. 2)

Without knowing the implications of these two questions, my friend successfully dodged the second one for he never gave an answer to it or probably he feigned ignorance.

"Why those questions?" I eventually asked him.
"I don't know," he answered.
"Well, I have some ideas and I would share them with you if you want me to."
"Go ahead," he said.

5.9 The A Word

"Well, I think your age is the issue here," I told him. Mr. Smith wanted to know your age and he knew that putting that to you directly as a question could be embarrassing. He therefore, decided to use a more subtle, mathematical and logical approach. To him, the answers to those two questions were all he needed to answer the third question he had in mind (What's your age?).

Take for example, if you lived y years in the United States, (see question no. 1) and x years in Nigeria (see question no. 2), then one doesn't need an adding machine to figure out what your age was."

In other words, let answer to question no. 1 $= y$ and answer to question no. 2 $= x$, and if your age = z, then $x + y = z$.

"That's simple arithmetic. But . . ."

5.10 Only One Assumption at the Moment

"There is of course, only one assumption, and that assumption is a restriction,"

I continued. "The assumption is that you have not lived anywhere else except United States, and Nigeria, your place of birth. That is the assumption and that is the restriction."

In other words, my friend's age is $x+y+r$ where
x = number of years he lived in Nigeria.
y = number of years he lived in the United States.
r = total number of years he has lived elsewhere, United States and Nigeria excluded.

If $r = 0$, his age is $x + y$.
Otherwise, his age would have been $x + y + r$, $r \geq 0$.
He stood motionless gazing at me.
I saw clearly his face saying:
"You are right."

5.11 Taking the Previous Assumption to the Next Level

Let us assume my friend lived:

1 year in the Soviet Union as a student.
1 year in Canada as a diplomat.
1 year in Kosovo as a soldier.
1 year in Iraq as a UN chemical inspection expert.
6 months in North Korea on a diplomatic mission.
30 years in Nigeria, his birthplace.
8 years in the United States as a resident.

From above, $x = 30$, $y = 8$,
$r = (1 + 1 + 1 + 1 + 0.5) = 4.5$.

Then, it will not take a rocket scientist to figure out that my friend's age is $(30 + 8 + 4.5) = 42.5$.

If he had not lived in any of the above foreign countries except the United States, then his age would have been $(30 + 8) = 38$, since $r = 0$. What do you think?

5.12 Transforming Word Problems into Simple Equations

Let's see how much you can transform these into equations.
Answers appear at the end of the questions.

1. Ten years ago Jim was 3 times as old as his son. Find the present age of the son if the sum of their ages ten years ago is 40 (**Ans:** 20).

2. The ages of Mrs. Smith's three children are consecutive odd numbers. If the youngest is y years old, how old is the oldest? (**Ans:** $y + 4$).

3. Gill has saved p dollars toward the purchase of a \$90.00 toy. How much more must he save to buy the toy? (**Ans:** 90-p dollars).

4. (a) Mr. Clark is now twice as old as his son who is d years old. How old was each of them five years ago? (**Ans:** Son: d - 5, Mr. Clark: 2d - 5) (b) If the sum of their present ages is 36, how old is Mr. Clark? (**Ans:** 24).

5. The length y of a rectangular field is 4 meters more than the width. Find the area A of the field (**Ans:** $A = y(y-4)$ sq. meters).

6. Two cylindrical tanks A and B of equal heights have radii 4 meters and 9 meters respectively. What is the difference in volume between A and B? (**Ans:** 65ph cubic meters).

7. The cost C in \$ of a helicopter wheel is given by $C = T + 2\text{p}r$ where T= cost in \$ for transporting one wheel from place of manufacture to point of sale and r = radius of wheel in centimeters. Find the cost of the wheel if $r = 280$ mm (**Ans:** C = T + 176).

8. Two cars A and B are traveling in the same direction at 30 mph and 45 mph respectively. If T represents time in hours, in how many hours are they P miles apart from each other? (**Ans:** $T = \frac{P}{15}$ hours)

9. The minimum and maximum speeds (in km per hour) posted on a Nigerian highway are a and b respectively. If their sum is y and their difference is 20, find the minimum and maximum speeds. (**Ans:** Minimum speed = $\frac{y-20}{2}$ km per hour. Maximum speed = $\frac{y+20}{2}$ km per hour)

10. A car consuming petrol at the rate of 10 km per liter at the cost of N600 per liter covered a distance of p kilometers. How much petrol did the car consume for the journey? (**Ans:** 60p naira).

11. A car traveled at v kilometers per hour for 2 hours and spent 30 minutes mending a puncture in the tire. Find the average speed of the car for the whole journey? (**Ans:** $\frac{2v}{5}$ km/h)

12. A school bought the following for its chemistry laboratory:

    5 conical flasks at p naira each
    9 troughs at q naira each
    Find the average cost of the items (**Ans:** $\frac{5p+9q}{14}$ naira).

13. The ages of 5 students are 15, 14, 13, 12, d. If the average age is x, find the age of the fifth student (**Ans:** d = 5x - 54).

14. Two cars A and B are travelling in opposite direction at 30 mph and 45 mph respectively. If T represents time in hours, in how many hours are they P miles apart from each other? (**Ans:** T = P/75 hours)

15. At a benefit concert, a Nigerian-US based non-profit organization raised the sum of T dollars for breast cancer victims in a remote

village in Nasarawa state. How much is this in naira if \$1.00 = N144 ? (**Ans:** 144T naira).

16. A triangle of height h cm has the same area with a rhombus whose base is 40 cm and height 20 cm. Find the base b of the triangle **(Ans:** $b = \frac{1600}{h}$ **cm).**

17. If the Celsius C and Fahrenheit F scales of temperature are connected by the relationship, $\frac{C}{5} = \frac{F-32}{9}$, find F in terms of C. At what temperature are the two scales equal? **(Ans:** $F = \frac{9}{5}C + 32$ **,** -40°).

18. If the Celsius and Fahrenheit scales of temperature are connected by the relationship, $\frac{C}{5} = \frac{F-32}{9}$, find C in terms of F (**Ans:** $C = \frac{5}{9}F - \frac{160}{9}$**).**

5.13 Strategies for Transforming Word Problems into Simple Equations

Transforming word problems into equations is like trying to recall a dream. In recalling a dream, you have to let your mind go. Visualize the problem, understand the problem, identify the unknowns (if applicable) and choose your variables wisely, provide a solution and verify your solution.

5.14 The if Game

If it is a case of simultaneous equation, solve the problem by substitution, and verify your solution using elimination method and vice versa.

If it is a case of clearing fractions, do so very carefully where necessary.

If it is a case of comparing coefficients and constant terms, do you know your coefficients?

Do you know your constant terms?
Do you know how to generate the equation(s) you need?

If it is a case of difference of two cubes, do you know the factors of $a^3 - b^3$?

$$a^3 - b^3 = (a-b)(a^2 + ab + b^2)$$

If it is a case of sum of two cubes, do you know the factors of $a^3 + b^3$?

$$a^3 + b^3 = (a+b)(a^2 - ab + b^2)$$

If it is a case of relative speeds involving cars, bicycles, trains etc. traveling in opposite directions, do you know that you have to add the speeds to get the relative speed?

If it is a case of relative speeds involving cars, bicycles, trains etc. traveling in the same direction, do you know that you have to subtract the speeds to get the relative speed?

If it is a case of direct or inverse proportion, find the constant of proportionality. From here can you use the principle of equivalent fractions comfortably as a computational tool?

If it is a case of using the cosine rule, can you identify the included angles? If it is a case of using the sine rule to solve a triangle completely, are all the necessary elements present?

Otherwise, can you determine the missing elements?

If it is a case of area of a triangle, do you know when to use Hero's formula? or $\frac{1}{2}(Base \times Height)$ or when to involve sine or cosine formula to find other elements of a triangle?

If it is a case of Pythagorean rule or theorem, can you identify the hypotenuse and apply the rule correctly?
If it is a case of comparing fractions, either ascending or descending order, do you know when to multiply by the LCM to find the least common

denominator (LCD)? It is easier to compare fractions with common denominators than fractions with different denominators.

If it is a case of simultaneous quadratic equation, do you know how to eliminate the unknowns? You can solve with the elimination method and verify using the substitution method. We can go on and on. The list is endless.

WORKED EXAMPLES

EXAMPLE 1: The sum of two multiples of 3 is 27. If the difference between twice the smaller and the larger is 9, find the numbers.

Using Simultaneous Equation

SOLUTION Since both numbers are multiples of 3, each can be expressed as a number $3a$ or $3b$.

If the smaller number $= 3a$ and the larger number $= 3b$, then
$3a + 3b = 27$ (sum of both numbers)

$2(3a) - 3b = 9$ (difference between twice the smaller, and the larger number)

$3a + 3b = 27$ ....(i)

$6a - 3b = 9$ ....(ii)

By adding equations (i) and (ii), $9a = 36 \Longleftrightarrow a = \frac{36}{9} = 4$.

If $3a + 3b = 27$, then $a + b = 9$ . . . (iii) (dividing both sides of the equation by 3).

By substituting for a in equation (iii) we have:
$4 + b = 9 \Longleftrightarrow b = 9 - 4 = 5$.

Therefore, if a = 4, and b = 5, then 3a =12 and 3b = 15.
The numbers are 12 and 15.

Using Expanded Notation

Let the smaller number = $cd = 10c + d$.
Let the larger number = $ab = 10a + b$.

Then $ab + cd = 27$ and $2(10c + d) - 10a - b = 9$.
$(10a + b) + (10c + d) = 27$ ....(i)
$20c + 2d - 10a - b = 9$..(ii)

By subtracting equation (ii) from equation (i) we have:
$10a + b + 10c + d - 20c - 2d + 10a + b = 27 - 9 = 18$

Simplifying the above by collecting like terms we have:
$(10a + 10a) + (b + b) + (10c - 20c) + (d - 2d) = 18$
$20a + 2b - 10c - d = 18$..(iii)

By adding equations (i) and (iii) we have
$(20a + 2b - 10c - d) + [(10a + b) + (10c + d)] = 18 + 27 = 45$

Simplifying the above by collecting like terms we have:
$(20a + 10a) + (2b + b) + (10c - 10c) + (d - d) = 45$
$30a + 3b = 45$ ....(iv)

Dividing both sides of equation (iv) by 3 we have:

$\frac{30a}{3} + \frac{3b}{3} = 10a + b = 15$.

If $(10a + b) + (10c + d) = 27$ and 10a + b = 15.

then by substitution, $10c + d = 27 - (10a + b) = 27 - 15 = 12$
Therefore, 10c + d =12 and $ab = 15$ and $cd = 12$
since a = 1, b = 5, c = 1, d = 2

Answers: Smaller number = 12 and the larger number = 15.

5.15 A Search for Counter Examples

Can you identify a temperature conversion situation such that

$$\frac{C}{5} \neq \frac{F-32}{9}$$

where F = temperature reading in Fahrenheit degrees and C = temperature reading in Centigrade degrees?

Can you identify a situation where the volume of a cylindrical tank (V) cannot be found using the relation, $V = \text{p}r^2h$?

Can you identify any two equivalent fractions $\frac{a}{b}$ and $\frac{c}{d}$ such that

$\frac{a}{b} \neq \frac{c}{d}$ or $\frac{a}{c} \neq \frac{b}{d}$ or $ad \neq bc$ or $\frac{c}{a} \neq \frac{d}{b}$?

Chapter 6

The Amazing Synchronicity

6.1 Objectives

At the end of the lesson, the students should be able to:

(a) Derive a rule for summing up first n positive odd integers.
(b) Discover the pattern in the sequence: 0, 1, 64, 243, 256, 125, ?, 7, 1.
(c) Supply the missing number in the sequence: 0, 1, 64, 243, 256, 125, ?, 7, 1.

6.2 Introduction

It was at Tecumseh High School, Tecumseh, a town about 35 miles east of Highway 9 near Norman, in the state of Oklahoma. I had just finished presenting a part of my *Math-Magic* workshop to a class of about 40 students. During the 50 minutes presentation, among other things, we derived a rule for finding the sum of first n positive odd numbers. In other words, we derived a rule for summing up n terms of the sequence: 1, 3, 5, 7, 9, 11, 13, 15, 17, 19, 21, 23, 25, 27, 29, 31 . . . In this chapter, we will derive the same rule.

6.3 Sum of First *n* Positive Odd Integers

We will start by decomposing the odd numbers as sums of 1's and 2's. Here are examples with three, four, and five such terms.

An Example with Three Terms

EXAMPLE 1: Consider 1 + 3 + 5..

1 =1
3 = 1 + (2)
5 = 1 + (2+2)
1 + 3 + 5 = 1 + (1+2)+(1+2+2)
= (1+1+1) + (2+2+2)
= 3 + (2+2+2)
= 3+3(2) = 3+3(3-1)..(i)

Let $3 = n$.

Now substituting for 3 in equation (i we have:
$3+3(3-1) = n+n(n-1) = n+n^2-n = n^2$.

Therefore, the sum of first *n* positive odd integers is n^2.

An Example with Four Terms

EXAMPLE 2: Let us consider 1 + 3 + 5 + 7.

1=1
3=1 + (2)
5= 1+(2+2)
7= 1+(2+2+2)

Now 1+3+5+7 = 1+(1+2)+(1+2+2)+(1+2+2+2)
= (1+1+1+1) + (2+2+2+2+2+2)

$= 4+(2+2)+(2+2)+(2+2)$
$= 4+(4+4+4) = 4+4(3) = 4+4(4-1)$ ....(i)

Let 4 = n.

Now substituting for 4 in equation (iv) we have:

$4+4(4-1) = n + n(n-1).$
$= n + n^2 - n = n^2.$

Therefore, the sum of first *n* positive odd integers is n^2.

An Example with Five Terms

EXAMPLE 3: Let us also consider 1+3+5+7+9.

$1 = 1$
$3 = 1 + (2)$
$5 = 1+(2+2)$
$7 = 1+(2+2+2)$
$9 = 1+ (1+2) + (1+2+2) + (1+2+2+2) + (1+2+2+2+2)$

Now $1+3+5+7+9 = 1+(1+2)+(1+2+2)+(1+2+2+2)+1+2+2+2+2)$
$= (1+1+1+1+1) + (2+2+2+2+2+2+2+2+2+2)$
$= 5+(2+2)+(2+2)+(2+2)+(2+2)+(2+2)$
$=5+(4+4+4+4+4) = 5+5(4) = 5+5(5-1)$...(i)

Let $5 = n$.

Now substituting for 5 in equation (i) we have:
$5+5(5-1) = n + n(n-1)$
$= n + n^2 - n = n^2.$

Therefore, the sum of first *n* positive odd integers is n^2.

6.4 Proof of Formula by Mathematical Induction

Let $1+3+5+7+9+\cdots+2n-1=n^2$
$P(1)=(1)^2=1$.

Therefore, P(1) is true since $(1)^2=1$.

If $P(n) = 1 + 3 + 5 + 7 + 9 + \ldots + (2n - 1) = n^2$

Then $P(n + 1) = (1 + 3 + 5 + 7 + 9 + \ldots + 2n - 1) + (2n + 1)$

$= n^2 + 2n + 1$ ....(i)
$P(n + 1) = (n + 1)^2 = n^2 + 2n + 1$ ....(ii)

Since (i) and (ii) are equal, $P(n+1)$ is also true.

6.5 A Search for Counter Examples

Can you identify any first n positive odd integers whose sum is not equal to n^2?

6.6 What a Coincidence!

When we finished, a student walked up to me and handed me a sheet of paper containing five problems titled, "October Problems of the Month" (given to the students by their mathematics teacher). Out of these five problems, I found one of them very interesting and relevant to what we investigated during the workshop. This was what I think motivated this student to initiate the contact.. Here is the problem.

Find and describe the pattern, then supply the missing number in the sequence:

0, 1, 64, 243, 256, 125, ?, 7, 1.

SOLUTION: **First Part of the Question**

Let ? = p. Then 0, 1, 64, 243, 256, 125, *p*, 7, 1
can also be written as 0^8, 1^7, $(2^2)^4$, 3^5, $(4^2)^2$, 5^3, *p*, 7^1, 1^0

$$0^8, 1^7, \left(2^2\right)^3, 3^5, \left(4^2\right)^2, 5^3, p, 7^1, 1^0$$

$$= 0^8, 1^7, 2^6, 3^5, 4^4, 5^3, p, 7^1, 8^0$$

The sequence could be described a^b
where $a = \{0,1,2,3,4,5,6,7,8\}$ and
$b = \{8,7,6,5,4,3,2,1,0\}$.

Second Part of the Question
By inspection, $a = 6$ which implies that $p = 6$, and $b = 2$.

Alternatively,
Let $p = a^b$ **= 7th term,** $b = L = a + (n-1)d$.

By substitution, $L = 8 + (7-1)(-1) = 8 + (-6) = 2$.
L = a + (n-1)d.

Again, by substitution, $L = 0 + (7-1)(1) = 6$.
$b = 2$. Therefore, $a^b = 6^2$.

Consequently, $a^b = 6^2 = p$.

Therefore, $p = 6^2$.

EXAMPLE 1 In a military school, student lockers are assigned odd numbers starting with 3. If the last locker is assigned the number 3025, how many lockers are there in the school?

SOLUTION: $L = a + (n-1)d$.

By substitution, $3025 = 3 + (n-1)d = 3 + (n-1)2$.
$3025 = 3 + 2n - 2 = 2n + 1$.

Therefore, $2n = 3024$.
From here, $n = 3024 \div 2 = 1512$.
Therefore, there are 1512 lockers in the school.

EXAMPLE 2 Prove that for *n* positive odd integers,

$$\frac{n}{2}(a + l) = n^2 + 2n$$

if and only if a = 3.

SOLUTION: $\frac{n}{2}(a + L) = \frac{n}{2}[a + a + (n - 1)\,d]$

$$\tfrac{n}{2}[a + a + 2n - 2] = \tfrac{n}{2}[2a + 2n - 2] = \tfrac{n}{2}(2)[a + n - 1] = n^2 + 2n$$

$$n(a + n - 1) = n(n + 2).$$

By comparing equal terms, $a + n - 1 = n + 2$, since $n = n$.
Transposing terms we have:

$$a = n + 2 - n + 1 = (n - n) + (2 + 1) = 3.$$

Therefore, $a = 3$. So, for *n* positive odd integers,
$\frac{n}{2}(a + l) = n^2 + 2n$ if and only if $a = 3$.

EXAMPLE 1 Consider 3, 5, 7, 9, . . ., 13.

$$n = \tfrac{L-a+2}{2} = \tfrac{13-3+2}{2} = 6$$

$a = 3$, $L = 13$, $n = 6$.

By substitution, $\frac{n}{2}(a + L) = \frac{6}{2}(3 + 13) = 3 \times 16 = 48$ ....(i)

By substitution, $n^2 + 2n = (6)^2 + (2 \times 6) = 36 + 12 = 48$ ....(ii)
Since (i) and (ii) above are equal, for n positive odd integers,

$\frac{n}{2}(a+l) = n^2 + 2n$ if and only if a = 3.

EXAMPLE 2 Consider the numbers 3, 5, 7, 9, . . ., 21.

$$n = \frac{L-a+2}{2} = \frac{21-3+2}{2} = 10$$

$a = 3$, $L = 21$, n = 10.

By substitution, $\frac{n}{2}(a+L) = \frac{10}{2}(3+21) = 5 \times 24 = 120$ ....(i)

By substitution, $n^2 + 2n = (10)^2 + (2 \times 10) = 100 + 20 = 120$ ....(ii)
Since (i) and (ii) above are equal, for n positive odd integers,

$\frac{n}{2}(a+l) = n^2 + 2n$ if and only if a = 3.

EXAMPLE 3 Consider the numbers 3, 5, 7, 9, . . . , 3025.

$$n = \frac{L- a+ 2}{2} = \frac{3025- 3+ 2}{2} = 1512$$

$a = 3, L = 3025, n = 1512$.

By substitution, $\frac{n}{2}(a+L) = \frac{1512}{2}(3+3025) = 756 \times 3028 = 2,289,168$ ....(i)

By substitution,

$n^2 + 2n = (1512)^2 + (2 \times 1512) = 2286144 + 3024 = 2,289,168$ ....(ii)

Since (i) and (ii) above are equal, for n positive odd integers,

$\frac{n}{2}(a+l) = n^2 + 2n$ if and only if a = 3.

EXAMPLE 4

Consider the numbers 3, 5, 7, 9, . . ., 4041.

$$n = \frac{L - a + 2}{2} = \frac{4041 - 3 + 2}{2} = 2020$$

$a = 3, L = 4041, n = 2020$.

By substitution, $\frac{n}{2}(a+L) = \frac{2020}{2}(3+4041) = 1010 \times 4044 = 4084440$ ....(i) if and only if $a = 3$.

By substitution,

$n^2 + 2n = (2020)^2 + (2 \times 2020) = 4080400 + 4040 = 4084440$.............(ii)

Since (i) and (ii) above are equal, for n positive odd integers,

$\frac{n}{2}(a+L) = n^2 + 2n$ if and only if $a = 3$.

EXAMPLE 5

Consider 3, 5, 7, 9 . . . 7221.

$$n = \frac{L - a + 2}{2} = \frac{7221 - 3 + 2}{2} = 3610$$

$a = 3,\ L = 7221,\ n = 3610$

By substitution, $\frac{n}{2}(a+L) = \frac{3610}{2}(3+7221) = 1805 \times 7224 = 13{,}039{,}320$ .. (i)

By substitution,
$n^2 + 2n = (3610)^2 + (2 \times 3610) = 13032100 + 7220 = 13039320$ ....(ii)

Since (i) and (ii) above are equal, for n positive odd integers,

$$\frac{n}{2}(a+L) = n^2 + 2n$$

if and only if a = 3.

EXAMPLE 6 Consider the numbers 3, 5, 7, 9 . . . 8243.

$$n = \frac{L - a + 2}{2} = \frac{8243 - 3 + 2}{2} = 4121$$

$a = 3, L = 8243, n = 4121$.

By substitution, $\frac{n}{2}(a + L) = \frac{4121}{2}(3 + 8243) = \frac{4121 \times 8246}{2} = 16990883$ ....(i)

By substitution
$n^2 + 2n = (4121)^2 + (2 \times 4121) = 16982641 + 8242 = 16990883$ ....(ii)

Since (i) and (ii) above are equal, for n positive odd integers,

$\frac{n}{2}(a + l) = n^2 + 2n$ if and only if a = 3.

6.7 Written Exercises

1. For the first n positive odd integers, the sum is given by n^2 and is also equal to $\frac{n}{2}(a + L)$. Derive the fact that a-1= 0.
2. Prove that for first n positive odd integers, $\frac{n}{2}(a + l) = n^2$, if a - 1 = 0.
3. Prove that a - 1 = 2 for first n positive odd integers whose first term is 3.

6.8 A Search for Counter Examples

(a) Can you identify any subset of consecutive positive odd integers whose first term is 3 such that for n terms, the sum is not equal to $n^2 + 2n$?
(b) Can you identify any subset of consecutive positive odd integers whose sum of n terms is equal to $n^2 + 2n$ but whose first term is not equal to 3?
(c) Can you identify first n positive odd integers such that the sum is not equal to n^2?

Chapter 7

Amazing Elementary Fibonacci Problems

7.1 Objectives

At the end of the lesson, the students should be able to:

a) Study worked problems based on Fibonacci sequence.
b) Prove some Fibonacci identities.
c) Use examples to verify some Fibonacci identities.
d) Prove that by contradiction, in the equation, $ax^2 + bx + c = 0$, the sum and product of roots will never be equal.

7.2 Introduction

In this chapter, *a*, *b*, *c*, *d*, and *e*, represent consecutive Fibonacci Numbers unless stated otherwise. This chapter will look at a few Fibonacci problems and their solutions. Six such problems are considered. Also treated here include verification of some Fibonacci identities. Proofs involving Fibonacci Numbers are also included. Such proofs are validated using Fibonacci identities.

7.3 Elementary Fibonacci Problems

A. Express $c^2 - 2ac + a^2$ as a single quantity

SOLUTION

But $c - a = b$.
Squaring both sides, we have: $(c-a)^2 = b^2$.
Therefore, $c^2 - 2ac + a^2 = b^2$.

B. Given that $a^2 + 2ab + b^2 = c^2$, express c as a sum of two Fibonacci Numbers

SOLUTION Given that $a^2 + 2ab + b^2 = c^2$.

Taking the square root of both sides, we have:

$$\sqrt{(a+b)^2} = \sqrt{c^2}$$

But $\sqrt{a^2 + 2ab + b^2} = \sqrt{(a+b)^2} = \sqrt{c^2}$.

From here, $c = a + b$.

Validating Our Result

First Validation We can validate our result from the fact that $a^2 + c^2 - 3ac + 1 = 0$.

The challenge and a part of the validation is to find the value of c in terms of a and b.

Given that $a^2 + c^2 - 3ac + 1 = 0$

$$a^2 + (a^2 + 2ab + b^2) - 3ac + 1 = 0$$

$$a^2 + b^2 + 2ab + a^2 + 1 = 3ac$$

From here, $c = \dfrac{a^2 + a^2 + b^2 + 2ab + 1}{3a}$

Simplifying further, we have:

$$c = \frac{2a^2 + b^2 + 2ab + ac - b^2}{3a}$$

$$\frac{2a^2 + b^2 + 2ab + a(b+a) - b^2}{3a}$$

$$= \frac{2a^2 + b^2 + 2ab + ab + a^2 - b^2}{3a}$$

$$\frac{3a^2 + 3ab}{3a^2} = \frac{3a(a+b)}{3a} = a + b$$

Therefore, $c = a + b$.

Another Perspective

REMARK: In validating our result in problem B, we found that

$$a^2 + c^2 - 3ac + 1 = \frac{2a^2 + b^2 + 2ab + 1}{3a}$$

and we also eventually from here, found that $c = a + b$.
We can show that

$$\frac{2a^2 + b^2 + 2ab + 1}{3a} = a + b$$

The logic is that if we can show that

$$\frac{2a^2 + b^2 + 2ab + 1}{3a} = a + b$$

then we have successfully showed that $c = a + b$, since $a + b = c$.

Given that $\frac{2a^2 + b^2 + 2ab + 1}{3a} = \frac{2a^2 + b^2 + 2ab - b^2 + ac}{3a}$

$$\frac{2a^2 + b^2 + 2ab - b^2 + ac}{3a} = \frac{2a^2 + (b^2 - b^2) + 2ab + ac}{3a}$$

$$= \frac{2a^2 + 2ab + a(a+b)}{3a}$$

$$\frac{2a^2 + 2ab + a^2 + ab}{3a} = \frac{3a^2 + 3ab}{3a}$$

$$\frac{3a^2 + 3ab}{3a} = \frac{3a(a+b)}{3a} = a + b.$$

But $a + b = c$ for any three consecutive Fibonacci Numbers a, b, and c. Taking the square of a+b we have:

$(a+b)^2 = a^2 + 2ab + b^2$ .... Equation 1

Therefore, $\frac{3a(a+b)}{3a} = \sqrt{2a^2 + 3ab - 1}$

By substitution, $2a^2 + 3ab - 1 = 2a^2 + 3ab + b^2 - ac$

$= 2a^2 + 3ab + b^2 - a(a+b)$

$2a^2 + 3ab + b^2 - a(a+b) = 2a^2 + 3ab + b^2 - a^2 - ab$

$= 2a^2 + (3ab - ab) + b^2 = a^2 + 2ab + b^2$ .... Equation 2

Since equation 2 is equal to equation 1, then

$$\frac{2a^2 + b^2 + 2ab + 1}{3a} = \sqrt{2a^2 + 3ab - 1}$$

for any three consecutive Fibonacci Numbers a, b, c where a and c have odd subscripts.

Validating Our Result with Examples

Example 1: If $a = 1$, $b = 1$, $c = 2$, then

$$\frac{2a^2 + b^2 + 2ab + 1}{3a}$$

$$= \frac{2(1)^2 + (1)^2 + (2 \times 1 \times 1) + 1}{3 \times 1}$$

$$= \frac{2 + 1 + 2 + 1}{3}$$

$$= \frac{6}{2} = 3$$

Therefore, c = 2 and 1 + 1 = 2.

EXAMPLE 2: If a = 5, b = 8 and c = 13.

$$\frac{2a^2 + b^2 + 2ab + 1}{3a}$$

$$= \frac{2(5)^2 + (8)^2 + (2 \times 5 \times 8) + 1}{3 \times 5}$$

$$= \frac{50 + 64 + 81}{15}$$

$$=\frac{195}{15}=13$$

Therefore, c = 13 and 5 + 8 = 13.

EXAMPLE 3: If a = 13, b = 21, and c =34, then

$$\frac{2a^2+b^2+2ab+1}{3a}$$

$$=\frac{2(13)^2+(21)^2+(2\times 13\times 21)+1}{3\times 13}$$

$$=\frac{338+441+547}{39}$$

$$=\frac{1326}{239}=34$$

Therefore, c = 34 and 13 + 21 = 34.

EXAMPLE 4: If a =34, b = 55 and c =89.

$$\frac{2a^2+b^2+2ab+1}{3a}$$

$$=\frac{2(34)^2+(55)^2+(2\times 34\times 55)+1}{3\times 34}$$

$$=\frac{2312+3025+3741}{102}$$

$$=\frac{9078}{102}=89$$

Therefore, c = 89 and 34 + 55 = 89.

EXAMPLE 5: If a =233, b = 377 and c = 610.

$$\frac{2a^2 + b^2 + 2ab + 1}{3a}$$

$$= \frac{2(233)^2 + (377)^2 + (2 \times 233 \times 377) + 1}{3 \times 233}$$

$$= \frac{108578 + 142129 + 1756831}{699}$$

$$= \frac{426390}{699} = 610$$

Therefore, c = 610 and 233 + 377 = 610.

EXAMPLE 6: If a = 610, b = 987 and c = 1597.

$$\frac{2a^2 + b^2 + 2ab + 1}{3a}$$

$$= \frac{2(610)^2 + (987)^2 + (2 \times 610 \times 987) + 1}{3 \times 610}$$

$$= \frac{744200 + 974169 + 1204210}{1830}$$

$$= \frac{2922510}{1830} = 1597$$

Therefore, c = 1597 and 610 + 987 = 1597.

REMARK: For any three consecutive Fibonacci Numbers a, b, c with odd subscripts, the following are true:

(a) $c = \sqrt{2a^2 + 3ab - 1}$

(b) $c = \dfrac{2a^2 + b^2 + 2ab + 1}{3a}$

Showing that $\dfrac{2a^2 + b^2 + 2ab + 1}{3a} = \sqrt{2a^2 + 3ab - 1}$

To show that the two values of *c* are equal, we can do this by using the principle of equivalent fractions which states that if two fractions $\frac{p}{q}$ and $\frac{r}{s}$ are equal, then

$p \times s = q \times r$ where $q \neq 0$ and $s \neq 0$

It is not only that applying the principle is mathematically correct but it is also a logical thing to do. We will do that and see what happens. Applying the same principle to our values of *c* we have:

$$\frac{2a^2 + b^2 + 2ab + 1}{3a} \equiv 2a^2 + 3ab - 1$$

$$\frac{2a^2 + b^2 + 2ab + 1}{3a} \equiv \frac{\sqrt{2a^2 + 3ab - 1}}{1}$$

By cross multiplication, $(1)(2a^2 + b^2 + 2ab + 1) \equiv 3a(\sqrt{2a^2 + 3ab - 1})$.

Taking the square of both sides, we have:

$$(2a^2 + b^2 + 2ab + 1)^2 \equiv 9a^2(2a^2 + 3ab - 1)$$

(Taking the square eliminates the radical sign.)

We have to expand the expression on the left hand side of the identity and see if it is equal to the expression on the right hand side.

Doing so we have: $(2a^2+b^2+2ab+1)^2$

$$=(2a^2+b^2+2ab+1)(2a^2+b^2+2ab+1)$$

$$=2a^2(2a^2+b^2+2ab+1)+b^2(2a^2+b^2+2ab+1)+2ab(2a^2+b^2+2ab+1)+1(2a^2+b^2+2ab+1)$$

$$=(4a^4+2a^2b^2+4a^3b+2a^2)+(2a^2b^2+b^4+2ab^3+b^2)+(4a^3b+2ab^3+4a^2b^2+2ab)$$

$$+1(2a^2+b^2+2ab+1)$$

$$=4a^4+8a^3b+8a^2b^2+4a^2+4ab^3+2b^2+4ab+b^4+1.$$

You will agree with me that the algebra is becoming too messy.

We therefore, need a change of strategy.

Before we can prove that $c=\sqrt{2a^2+3ab-1}=\dfrac{2a^2+b^2+2ab+1}{3a}$

(under certain conditions), we have to prove that $c=\sqrt{2a^2+3ab-1}$

or that $c^2=2a^2+3ab-1$. Let us do exactly that.

Proof that $2a^2+3ab-1=c^2$ *or that* $c=\sqrt{2a^2+3ab-1}$

If $2a^2+3ab-1=c^2$, then by substitution,

$$2a^2+3ab-1=2a^2+3ab-(-b^2+ac)$$

$$=2a^2+3ab+b^2-ac$$

REMARK: For any three consecutive Fibonacci Numbers a, b, c with a and c odd subscripted,

$b^2-ac=-1$ and $(-1)(b^2-ac)=(-1)(-1)$.

$$2a^2+3ab+b^2-ac=2a^2+3a(c-a)+b^2-ac$$

$$=(2a^2-3a^2)+(3ac-ac)+b^2$$

$$=-a^2+2ac+b^2=b^2-a^2+2ac$$

$$=(b-a)(b+a)+2ac=(b-a)c+2ac$$

$$=bc-ac+2ac=bc+ac=c(b+a)=c^2.$$

(By definition, $c = a + b$ or $c = b + a$.)

Therefore, for any three consecutive Fibonacci Numbers a, b, c with a and c odd-subscripted,

$$c^2 = 2a^2 + 3ab - 1.$$

By taking square roots of both sides,

$$c = \sqrt{2a^2 + 3ab - 1}.$$

Proof that $\dfrac{2a^2 + b^2 + 2ab + 1}{3a} = \sqrt{2a^2 + 3ab - 1}$

If $\dfrac{2a^2 + b^2 + 2ab + 1}{3a} = \sqrt{2a^2 + 3ab - 1}$, then

$\dfrac{2a^2 + b^2 + 2ab + 1}{3a}$ should evaluate to c when simplified since

$$c = \sqrt{2a^2 + 3ab - 1}$$

Proceeding with the evaluation process we have:

$$\frac{2a^2+b^2+2ab+1}{3a} = \frac{2a^2+b^2+2ab-b^2+ac}{3a}$$

$$= \frac{2a^2+(c-a)^2+2a(c-a)-(c-a)^2+ac}{3a}$$

$$= \frac{2a^2+(c-a)^2+2a(c-a)^2+2a(c-a)+ac}{3a}$$

$$\frac{2a^2+2ac-2a^2+ac}{3a} = \frac{(2a^2-2a^2)+(2ac+ac)}{3a} = c.$$

This is yet another proof.

Verification Using Chain Arguments

There are many ways of finding c in terms of a and b as there are identities of the type listed below:

Fibonacci Identity	*Fibonacci Identity*
$ae - c^2 = \pm 1$	$be - cd = \pm 1$
$ad - bc = \pm 1$	$cd - be = \pm 1$
$ac - b^2 = \pm 1$	$c^2 - ac = \pm 1$
$b^2 - ac = \pm 1$	$be - ad = \pm 1$

Validating Our Result Using Fibonacci Identities

Below is another way we can validate our result using one of the Fibonacci identities.

$$a^2 + c^2 - 3ac + 1 = 0$$

$$a^2 + c^2 - 3ac + (be - cd)$$

By substituting for c and d we have:

$$a^2 + c^2 - 3ac - be + cd = 0$$

$$= a^2 + c^2 - 3ac - b(c + d) + d(a + b) = 0$$

$$a^2 + c^2 - 3ac - bc - bd + ad + bd = 0$$

$$= a^2 + c^2 - 3ac + ad - bc = 0$$

$$= a^2 + c^2 - 3ac + a(b + c) \ - b(a + b) = 0$$

$$= a^2 + c^2 - 3ac + a(b + c) - b(a + b) = 0$$

$$= a^2 + c^2 - 3ac + ab + ac - ab - b^2 = 0$$

$$= a^2 + c^2 - 3ac + ac - b^2 = 0$$

$$= a^2 + c^2 - 2ac - b^2 = 0$$

$$a^2 + (a+b)^2 - 2ac - b^2 = 0$$

$$= a^2 + (a^2 + 2ab + b^2) - 2ac - b^2 = 0$$

$$(a^2 + a^2) + 2ab + (b^2 - b^2) - 2ac = 0$$

$$2a^2 + 2ab = 2ac$$

$$\frac{2a(a+b)}{2a} = \frac{2ac}{2} \Leftrightarrow c = a + b$$

Therefore, $c = a + b$.

Second Validation Using Fibonacci Identities

This is another way of validating our result using Fibonacci identities.

$$a^2 + c^2 - 3ac + 1 = 0$$

$$a^2 + c^2 - 3ac + (ae - c^2) = 0$$

$$a^2 + c^2 - 3a(a+b) + [a(c+d) - c^2] = 0$$

$$a^2 + c^2 - 3a^2 - 3ab + ac + ad - c^2 = 0$$

$$(a^2 - 3a^2) - 3ab + ac + ad = 0$$

$$-2a^2 - 3ab + a(a+b) + a(b+c) = 0$$

$$-2a^2 - 3ab + a^2 + ab + ab + ac = 0$$

$$(-2a^2 + a^2) - (3ab + 2ab) + ac = 0$$

$$-a^2 - ab + ac = 0$$

$$a^2 + ab = ac$$

$$a(a + b) = ac$$

$$\frac{a(a+b)}{a} = \frac{ac}{a} \Leftrightarrow c = a + b$$

Third Validation Using Fibonacci Identities

$$a^2 + c^2 - 3ac + 1 = 0$$

$a^2 + c^2 - 3ac + ac - b^2 = 0$

$\{b^2 - ac = -1\}$ (Multiplying both sides by -1).

$a^2 + c^2 - 3ac + ac - b^2 = 0$

$a^2 + c^2 - 2ac - b^2 = 0$

$(a-c)^2 - b^2 = 0$

Factoring we have: $(a - c - b)(a - c + b) = 0$

Either a - c - b = o or a - c + b = 0 $\Leftrightarrow$ c = a - b or c = a + b.

But $c = a - b$ is unsuitable because it is inconsistent with one of the basic properties of the Fibonacci sequence that the nth term of the Fibonacci sequence is equal to the sum
of the two preceeding terms,

$$u_n + u_{n-1} = u_{n+1}, n \geq 1.$$

Therefore, $c = a + b$.

Fourth Validation Using Fibonacci Identities

$$a^2 + c^2 - 3ac + 1 = 0$$

$$(c-b)^2 + c^2 - 3ac + 1 = 0$$

$$b^2 - ac = -1 \Leftrightarrow -b^2 + ac = 1$$

$$(c-b)^2 + c^2 - 3ac + 1 = c^2 - 2bc + b^2 + c^2 - 3ac - b^2 + ac = 0$$

$$= (c^2 + c^2) - 2bc + (b^2 - b^2) - 3ac + ac = 0$$

$$2c^2 - 2bc - 2ac = 0$$

$$2ac = 2c^2 - 2bc$$

$$ac = c^2 - bc = c(c-b)$$

Dividing both sides by c we have:

$$\frac{ac}{c} = \frac{c(c-b)}{c} \Leftrightarrow a = c - b$$

From here, a = c-b and c = a+b.

C. ***Prove that for any five consecutive Fibonacci numbers a, b, c, d, e, ae - c^2 = ±1***

To prove that : $ae - c^2 = \pm 1$

Proof

$$ae - c^2 = a(c+d) - (a+b)^2 = ac + ad - a^2 - 2ab - b^2$$

$$= a(a+b) + a(b+c) - a^2 - 2ab - b^2$$

$= a^2 + ab + ab + ac - a^2 - 2ab - b^2$

$= (ab + ab) + (a^2 - a^2) - 2ab + ac - b^2$

$= 2ab + ac - 2ab - b^2 = ac - b^2$ and $ac - b^2 = \pm 1$.

Therefore, $ae - c^2 = \pm 1$.

Miscellaneous Examples

EXAMPLE 1

If $a = 1$, $b = 1$, $c = 2$, $d = 3$, $e = 5$,

$ae - c^2 = (1 \times 5) - 2^2$

$= 5 - 4 = 1$

EXAMPLE 2

If $a = 13, b = 21, c = 34, d = 55, e = 89,$

$ae - c^2 = (13 \times 89) - 34^2$

$= 1157 - 1156 = 1$

An Example with Three Digit Terms

If a = 144, b = 233, c = 377, d = 610, e = 987,

$ae - c^2 = (144 \times 987) - 377^2$

$= 142128 - 141129 = -1$

D. Prove that for any five consecutive Fibonacci numbers a, b, c, d, e, ad - bc = ±1

To prove that $ad - bc = \pm 1$

Proof

ad-bc = ± 1
= a(b+c) - b(a+b)
= (ab-ab) + $ac - b^2$
= $ac - b^2$

Therefore, ad-bc $= \pm 1$, since $ac - b^2 = \pm 1$, and $ac - b^2$=ad-bc.

EXAMPLE 1

When a = 2, b = 3, c = 5, d = 8,
ad-bc = (2 × 8) - (3 × 5)
= 16 - 15 = 1.

Therefore, ad-bc $= \pm 1$, since $ac - b^2 = \pm 1$, and $ac - b^2$=ad-bc.

EXAMPLE 2

When a = 13, b = 21, c = 34, d = 55,
ad-bc = (13 × 55) - (21 × 34)
= 715 - 714 = 1.

Therefore, ad-bc $= \pm 1$, since $ac - b^2 = \pm 1$, and $ac - b^2$=ad-bc.

EXAMPLE 3

When $a = 233$, $b = 377$, $c = 610$, $d = 987$,
$ad - bc = (233 \times 987) - (377 \times 610)$
229971-229970 =1

Therefore, ad-bc $= \pm 1$, since $ac - b^2 = \pm 1$, and $ac - b^2$=ad-bc.

EXAMPLE 4

When $a = 987$, $b = 1597$, $c = 2584$, $d = 4181$,

ad-bc$=(987\times 4181)-(1597\times 2584)$

4126647 - 4126648 = -1

Therefore, ad-bc $=\pm 1$, since $ac-b^2=\pm 1$, and $ac-b^2$=ad - bc.

E. Proof that for any five consecutive Fibonacci numbers a, b, c, d, e, be - cd = ±1

To Prove that $be-cd=\pm 1$

Proof: $be-cd=b(c+d)-c(b+c)$

$=bc+bd-bc-c^2 \; =bd-c^2$

$=b(b+c)-(a+b)^2 \; =b^2+bc-(a^2+2ab+b^2)$

$=b^2+bc-a^2-2ab-b^2$.

$=b^2+b(a+b)-a^2-2ab-b^2$

$=b^2-ab-a^2=(b^2-a^2)-ab$

$=(b+a)(b-a)-ab=c(b-a)-ab$

$=bc-ac-ab=b(a+b)-ac-ab$

$=ab+b^2-ac-ab=b^2-ac$

But b^2 - ac. =-1.

Therefore, be-cd = b^2 - ac.

Consequently, be-cd = ±1.

EXAMPLE 1

$a = 1, b = 1, c = 2, d = 3, e = 5$
$be - cd = (1\times5) - (2\times3) = 5 - 6 = -1$

EXAMPLE 2

$a = 2, b = 3, c = 5, d = 8, e = 13$
$be - cd = (3\times13)-(5\times8) = 39 - 40 = -1$

EXAMPLE 3

$a = 3, b = 5, c = 8, d = 13, e = 21$
be - cd = $(5\times21)-(8\times13) = 105-104 = 1$

F. Prove that for any four consecutive Fibonacci Numbers,

$a, b, c, d, ac - b^2 = ad - bc$

To prove that $ac - b^2 = ad - bc$

Proof

$$ac - b^2 = a(d - b) - b^2$$

$$ac - b^2 = a(d - b) - b^2$$

$$= a(d - b) - c^2 + 2ac - a^2$$

$$= ac - c^2 + 2ac - a^2$$

$$= c(a - c) + 2ac - a^2$$

$$= c(-b) + a(2c - a) = -cb + a(2c - a)$$

$$-cb + a[2(a + b) - a]$$

$= -cb + a[(2a+2b) - a]$

$= -cb + a[2a+2b - a]$

$= -cb + a^2 + 2ab$

$= -cb + (a^2 + ab) - ab = -cb + a^2 + ab + a(c - a)$

$-cb + a^2 + ab + ac - a^2 = -cb + (ab + ac)$

$= -cb + a(b + c) = -cb + ad = ad - bc$

Therefore, for any four consecutive Fibonacci numbers,

$a, b, c, d, ac - b^2 = ad - bc$

Examples with One Digit Terms

EXAMPLE 1

If $a = 1$, $b = 1$, $c = 2$, $d = 3$,

$ac - b^2 = (1 \times 2) - 1^2 = 2 - 1 = 1$

$ad - bc = (1 \times 3) - (1 \times 2) = 3 - 2 = 1$

Therefore, $ac - b^2 = ad - bc.$

EXAMPLE 2

If $a = 1$, $b = 2$, $c = 3$, $d = 5$,

$ac - b^2 = (1 \times 3) - 2^2 = 3 - 4 = -1$

$ad - bc = (1 \times 5) - (2 \times 3) = 5 - 6 = -1$

Therefore, $ac - b^2 = ad - bc.$

Examples with Two Digit Terms

EXAMPLE 1

If $a = 13, b = 21, c = 34, d = 55,$
$ac - b^2 = (13 \times 34) - 21^2$
$= 442 - 441 = 1$
$ad - bc = (13 \times 35) - (21 \times 34$
$= 715\text{-}714 = 1$
Therefore, $ac - b^2 = ad - bc.$

EXAMPLE 2

If $a = 21, b = 34, c = 55, d = 89,$
$ac - b^2 = (21 \times 55) - 34^2$
$= 1155 - 1156 = -1$
$ad - bc = (21 \times 89) - (34 \times 55)$
$= 1869\text{-}1870 = \text{-}1$
Therefore, $ac - b^2 = ad - bc.$

Examples with Three Digit Terms

EXAMPLE 1

If $a = 144, b = 233, c = 377, d = 610,$
$ac - b^2 = (144 \times 377) - 233^2$
$= 54288 - 54289 = -1$
$ad - bc = (144 \times 610) - (233 \times 377)$
$= 87840\text{-}87841 = \text{-}1$
Therefore, $ac - b^2 = ad - bc.$

EXAMPLE 2

If a = 233, b = 377, c = 610, d = 987,
$ac - b^2 = (233 \times 610) - 377^2$
$= 142130 - 142129 = 1$

$ad - bc = (233 \times 987) - (377 \times 610)$
$= 229971\text{-}229970 = 1$
Therefore, $ac - b^2 = ad - bc$.

G. Prove that cd - ab is a Fibonacci number if a,b, c, d are four consecutive Fibonacci numbers.

To prove that *cd - ab* is a Fibonacci number

Proof: cd-ab = c(b+c) - ab

$c(b + c) - ab = bc + c^2 - ab = c^2 + bc - ab = c^2 + b(c - a)$

$c^2 + b(c - a) = c^2 + b^2 = b^2 + c^2$

The sum of squares of two consecutive Fibonacci numbers b and c is a Fibonacci number whose subscript is equal to the sum of subscripts of b and c.

Therefore, $b^2 + c^2$ and consequently, cd-ab are Fibonacci numbers since cd-ab = $b^2 + c^2$.

H. Prove that in any quadratic equation, the sum of roots and product of roots will never be equal where a, b, c are three consecutive Fibonacci numbers.

Proof

The general form of a quadratic equation is given by

$ax^2 + bx + c = 0, a \neq 0$

If $ax^2 + bx + c = 0$, then

$x^2 - \frac{b}{a}x + \frac{c}{a} = 0$ *(dividing both sides by a)*

$$x^2 - \left\{\frac{c-a}{c-b}\right\}x + \frac{a+b}{c-b} = 0$$

Sum of roots = $\dfrac{c-a}{c-b}$

Product of roots = $\dfrac{a+b}{c-b}$

To prove that the sum of roots and product of roots will never be equal, we have to assume that they are and then prove otherwise by contradiction.

Doing so we have:

$$\frac{c-a}{c-b} \equiv \frac{a+b}{c-b} \Leftrightarrow (c-\ a)(c-\ b) = (c-\ b)(a+b)$$

If sum of roots and product of roots are equal, then

$$c - a = a + b \Leftrightarrow c = 2a + b$$

This is a contradiction since for any three consecutive Fibonacci numbers a, b, c

c=a+b and not 2a+b.

Therefore, the sum of roots and product of roots of the equation, $ax^2 + bx + c = 0$ will never be equal.

Another Perspective

If we assume that the sum of roots and product of roots are equal, then

$$\frac{c-a}{c-b} \equiv \frac{a+b}{c-b}$$

If we assume that $\frac{c-a}{c-b} \equiv \frac{a+b}{c-b}$, then we are also assuming that $\frac{c-a}{c-b}$ and $\frac{a+b}{c-b}$ are equivalent. Therefore, since by applying the principle of equivalent fractions, the denominators are equal, the numerators are also equal.

Therefore, $c-a=a+b$.

From here, c = 2a + b.

The expression, c = 2a+b is a contradiction, since for any three consecutive Fibonacci numbers a, b, c, c - a = a +b and not c = 2a+b.

Therefore, the sum of roots and product of roots of the equation, $ax^2 + bx + c = 0$ will never be equal.

Problem 5

Given that $b^2 - ac = 1$ and $a^2 + c^2 - 3ac - 1 = 0$, if and only if a and c have even subscripts, find a in terms of b and c

SOLUTION

$$a^2 - 3ac + c^2 = b^2 - ac \Leftrightarrow a^2 - 2ac + c^2 - b^2 = 0$$

$$= a^2 - 2ac + c^2 - b^2 \Leftrightarrow (a-c)^2 - b^2 = 0$$

Factorizing we have:

$$(a - c - b)(a - c + b) = 0$$

If $(a - c - b)(a - c + b) = 0$, then the following are true:

Either, $a - c - b = 0 \Leftrightarrow a = c + b$ or $a - c + b = 0 \Leftrightarrow a = c - b$

For any three consecutive Fibonacci numbers a, b, c, $c = a+b$ so that $a = c - b$ and not $a = c + b$.

Problem 6

Prove that $a_{n-1} + a_{n+1}$ is a multiple of 5 where a_n is the nth Lucas number and a_n is defined recursively as $u_{n-1} + u_{n+1}$, n^3 1 where u_n is the nth Fibonacci number.

Given $a_n = u_{n-1} + u_{n+1}, n \geq 1$

To prove that $a_{n-1} + a_{n+1}$ is a multiple of 5 where a_n is the nth Lucas number.

Proof

Re-writing $a_{n-1} + a_{n+1}$ we have:

$$a_{n-1} + a_{n+1} = a_{n-1} + (a_n + a_{n-1})$$

$$= a_{n-1} + \left(u_{n-1} + u_{n+1}\right) + \left(u_{n-2} + u_n\right)$$

$$= \left(u_{n-2} + u_n\right) + u_{n-1} + u_{n+1} + \left(u_{n-2} + u_n\right)$$

$$= u_{n-2} + u_{n-1} + u_n + u_{n+1} + u_{n-2} + u_n$$

$$= (u_{n-2} + u_{n-1}) + u_n + u_n + (u_n + u_{n-1}) + (u_n - u_{n-1})$$

$$= \left(u_n + u_n + u_n + u_n\right) + (u_{n-1} - u_{n-1}) + (u_{n-2} + u_{n-1})$$

$$= (u_n + u_n + u_n + u_n + u_n) + (u_{n-1} - u_{n-1})$$

$$= (u_n + u_n + u_n + u_n + u_n) = 5u_n$$

Since $5u_n = a_{n-1} + a_{n+1}$, 5 is a multiple of 5 and since $5u_n = a_{n-1} + a_{n+1}$, 5 is also a multiple of $a_{n-1} + a_{n+1}$.

7.4 Written Exercises

a) Prove that for any five consecutive Fibonacci numbers a, b, c, d, e, $bc - ad = \pm 1$.

b) Prove that for any five consecutive Fibonacci numbers a, b, c, d, e, $ce - bd = \pm 1$.

c) Prove that for any five consecutive Fibonacci numbers a, b, c, d, e, $b^2 - ac = \pm 1$.

7.4 A Search for Counter Examples

a) Can you identify any four consecutive Fibonacci numbers a, b, c, d such that $ac - b^2 \neq ad - bc$?

b) Can you identify any five consecutive Fibonacci numbers *a, b, c, d* such that $ae - c^2 \neq ad - bc$?

c) Can you identify any four consecutive Fibonacci numbers *a, b, c, d* such that $ad - bc \neq \pm 1$?

Chapter 8

Mathematics from Out of State

8.1 Objectives

At the end of the lesson, the students should be able to:

(a) Appreciate and understand the beautiful property of 89, the eleventh term of the Fibonacci sequence.

(b) Prove the following identities for any four consecutive Fibonacci numbers, $u_n, u_{n+1}, u_{n+2}, u_{n+3}$.

$$\frac{u_{n+3}}{2} + \frac{u_n}{2} = u_{n+2} \qquad\qquad \frac{u_{n+3}}{2} - \frac{u_n}{2} = u_{n+1}$$

(c) Identify any four consecutive Fibonacci numbers

$u_n, u_{n+1}, u_{n+2}, u_{n+3}$

such that $\frac{u_{n+3}}{2} + \frac{u_n}{2} {}^{1} \; u_{n+2}$

(b) Identify any four consecutive Fibonacci numbers

$u_n, u_{n+1}, u_{n+2}, u_{n+3}$

such that $\frac{u_{n+3}}{2} + \frac{u_n}{2}{}^{1} \quad u_{n+1}$

8.2 Introduction

My love and admiration for Fibonacci numbers started when I was a student at Alvan Ikoku College of Education, Owerri, Imo state, Nigeria. By then, I was the only student member of the National Council of Teachers of Mathematics (NCTM). When I immigrated to the United States, I decided to continue my affiliation with NCTM. In the January 6, 1996 issue of The Mathematics Teacher, a professional magazine of NCTM, a professor at Adams State College, Alamosa, Colorado, wrote an article on the Fibonacci sequence, titled, "Vol. 89 and 11 to the Centennial."

By then, I have started writing my book on the subject and I covered a part of the professor's topic. This got me excited and I decided to contribute in form of a rejoinder which now follows:

8.3 A Rejoinder to a Magazine Article

Joan Armstead
The Senior Journal Editor
The Mathematics Teacher
1906 Association Drive
Reston, VA 22091

May I refer to an interesting article by Monte J. Zerger titled, "Vol. 89 (and 11 to Centennial)" that appeared on p. 3 of January issue of *The Mathematics Teacher*. The said article pointed out some beautiful properties of the eleventh Fibonacci number 89. This may be an opportunity also to point out the amazing property of the only two-digit Fibonacci number which is related to that portion of the article.

To arrive at 1189 we can also consider the above processes, which is an adaptation from my book, *A Course in Fibonacci Numbers*. In the article, it was mentioned that the following properties are true of Fibonacci numbers.

$$(A)\frac{u_{n+3}}{2}+\frac{u_n}{2}=u_{n+2} \quad (B)\frac{u_{n+3}}{2}-\frac{u_n}{2}=u_{n+1}$$

Let u_n represent the nth Fibonacci number. In my book on the subject, I proved the above properties as follows:

8.3 Proof of a Fibonacci Identity, Part A

$$(A) \quad \frac{u_{n+3}}{2}+\frac{u_n}{2}=u_{n+2}$$

To prove that $\frac{u_{n+3}}{2}+\frac{u_n}{2}=u_{n+2}$.

Proof

$$\frac{u_{n+3}}{2}+\frac{u_n}{2} = \frac{1}{2}u_{n+3}+\frac{1}{2}u_n$$

$$=\left(\frac{1}{2}u_{n+1}+\frac{1}{2}u_{n+2}\right)+\left(\frac{1}{4}u_n+\frac{1}{4}u_n\right):$$

$$=\frac{1}{2}u_{n+2}+\frac{1}{2}u_{n+1}+\frac{1}{4}u_n+\frac{1}{4}u_n$$

$$=\left(\frac{1}{4}u_{n+2}+\frac{1}{4}u_{n+2}\right)+\left(\frac{1}{4}u_{n+1}+\frac{1}{4}u_{n+1}\right)+\left(\frac{1}{4}u_n+\frac{1}{4}u_n\right)$$

$$=\left(\frac{1}{4}u_n+\frac{1}{4}u_{n+1}\right)+\left(\frac{1}{4}u_n+\frac{1}{4}u_{n+1}\right)+\left(\frac{1}{4}u_{n+2}+\frac{1}{4}u_{n+2}\right)$$

$$=\frac{1}{4}(u_n+u_{n+1})+\frac{1}{4}(u_n+u_{n+1})+\frac{1}{2}(u_{n+2})$$

$$= \left(\frac{1}{4}u_{n+2} + \frac{1}{4}u_{n+2}\right) + \frac{1}{2}\left(u_{n+2}\right)$$

$$= \left(\frac{1}{2}u_{n+2} + \frac{1}{2}u_{n+2}\right) = u_{n+2}$$

Therefore, $\frac{u_{n+3}}{2} + \frac{u_n}{2} = u_{n+2}$

8.4 Illustrating with Examples

EXAMPLE 1

Let $5 = u_n$, $8 = u_{n+1}$, $13 = u_{n+2}$, and $21 = u_{n+3}$.

$$\frac{u_{n+3}}{2} + \frac{u_n}{2} = \frac{21}{2} + \frac{5}{2} = \frac{26}{2} = 13$$

and $u_{n+2} = 13$.

Therefore, $\frac{u_{n+3}}{2} + \frac{u_n}{2} = u_{n+2}$.

EXAMPLE 2

Let $8 = u_n$, $13 = u_{n+1}$, $21 = u_{n+2}$, and $34 = u_{n+3}$.

$$\frac{u_{n+3}}{2} + \frac{u_n}{2} = \frac{34}{2} + \frac{8}{2} = \frac{42}{2} = 21$$

and $u_{n+2} = 21$

Therefore, $\frac{u_{n+3}}{2} + \frac{u_n}{2} = u_{n+2}$.

EXAMPLE 3

Let $34 = u_n$, $55 = u_{n+1}$, $89 = u_{n+2}$, and $144 = u_{n+3}$.

$$\frac{u_{n+3}}{2}+\frac{u_n}{2} = \frac{144}{2}+\frac{34}{2} = \frac{178}{2} = 89$$

and $u_{n+2} = 89$.

Therefore, $\frac{u_{n+3}}{2}+\frac{u_n}{2} = u_{n+2}$.

EXAMPLE 4

Let $144 = u_n$, $233 = u_{n+1}$, $377 = u_{n+2}$, and $610 = u_{n+3}$.

$$\frac{u_{n+3}}{2}+\frac{u_n}{2} = \frac{610}{2}+\frac{144}{2} = \frac{754}{2} = 377.$$

and $u_{n+2} = 377$.

Therefore, $\frac{u_{n+3}}{2}+\frac{u_n}{2} = u_{n+2}$.

8.6 Proof of a Fibonacci Identity, Part B

$$(B)\ \frac{u_{n+3}}{2}-\frac{u_n}{2} = u_{n+1}$$

To prove that $\frac{u_{n+3}}{2}-\frac{u_n}{2} = u_{n+1}$.

Proof

$$\frac{u_{n+3}}{2}-\frac{u_n}{2} = \frac{(u_{n+2}+u_{n+1})}{2}-\frac{u_n}{2}$$

$$= \left(\frac{u_{n+2}}{2}+\frac{u_{n+1}}{2}\right)-\frac{u_n}{2}$$

$$= \left(\frac{1}{4}u_{n+2}+\frac{1}{4}u_{n+2}\right)+\left(\frac{1}{4}u_{n+1}+\frac{1}{4}u_{n+1}\right)-\left(\frac{1}{4}u_n+\frac{1}{4}u_n\right)$$

$$= \left(\frac{1}{4}u_{n+1}+\frac{1}{4}u_{n+1}\right)+\left(\frac{1}{4}u_{n+2}-\frac{1}{4}u_n\right)+\left(\frac{1}{4}u_{n+2}-\frac{1}{4}u_n\right)$$

$$= \left(\frac{1}{4}u_{n+1} + \frac{1}{4}u_{n+1}\right) + \frac{1}{4}u_{n+1} + \frac{1}{4}u_{n+1}$$

$$= \frac{1}{2}u_{n+1} + \left(\frac{1}{4}u_{n+1} + \frac{1}{4}u_{n+1}\right) = \frac{1}{2}u_{n+1} + \frac{1}{2}u_{n+1} = u_{n+1}$$

Therefore, $\frac{u_{n+3}}{2} - \frac{u_n}{2} = u_{n+2}$.

8.5 Illustrating with Examples

EXAMPLE 1

Let $5 = u_n$, $8 = u_{n+1}$, $13 = u_{n+2}$, and $21 = u_{n+3}$.

$$\frac{u_{n+3}}{2} - \frac{u_n}{2} = \frac{21}{2} - \frac{5}{2} = \frac{16}{2} = 8.$$

and $u_{n+1} = 8$.

Therefore, $\frac{u_{n+3}}{2} - \frac{u_n}{2} = u_{n+1}$.

EXAMPLE 2

Let $8 = u_n$, $13 = u_{n+1}$, $21 = u_{n+2}$, and $34 = u_{n+3}$.

$$\frac{u_{n+3}}{2} - \frac{u_n}{2} = \frac{34}{2} - \frac{8}{2} = \frac{26}{2} = 13.$$

and $u_{n+1} = 13$.

Therefore, $\frac{u_{n+3}}{2} - \frac{u_n}{2} = u_{n+1}$.

EXAMPLE 3

Let $34 = u_n$, $55 = u_{n+1}$, $89 = u_{n+2}$, and $144 = u_{n+3}$.

$$\frac{u_{n+3}}{2} - \frac{u_n}{2} = \frac{144}{2} - \frac{34}{2} = \frac{110}{2} = 55.$$

and $u_{n+1} = 55$.

Therefore, $\frac{u_{n+3}}{2} - \frac{u_n}{2} = u_{n+1}$.

EXAMPLE 4

Let $144 = u_n$, $233 = u_{n+1}$, $377 = u_{n+2}$, and $610 = u_{n+3}$.

$$\frac{u_{n+3}}{2} - \frac{u_n}{2} = \frac{610}{2} - \frac{144}{2} = \frac{466}{2} = 233.$$

and $u_{n+1} = 233$.

Therefore, $\frac{u_{n+3}}{2} - \frac{u_n}{2} = u_{n+1}$.

The article was submitted to the editor of *The Mathematics Teacher* who in turn sent it to a professor at Adams State College, Alamosa, CO for review. The professor's comments start right here.

8.6 Professor's Monte Zerger's Comments

Adam's State College
School of Science, Mathematics & Technology
Alamosa, CO 81102
March 8, 1996

Editor
Mathematics Teacher
1906 Association Drive
Reston, VA 22091

Dear Editor,

Below are some comments addressing Emekwulu's "Vol. 89 Revisited". I hope the first portion regarding 1189 can be published. I do not feel the same way about the second portion after the word, WOW. Since apparently, he did not include any explanation for these seemingly magical occurrences of 1189, perhaps one such as mine below should be included.

Also, let me point out a type in the 11th line which should read. For our new sequence, the following are true.

Comments: He went on to comment further as follows: Beautiful indeed! After some frustrating moments, I "got the drift." It hinges on the fact that the second sequence is still a "Fibonacci type sequence," a sequence in which the sum of any two successive terms yields the next term. In any such sequence, all expressions corresponding to Emekwulu's will have a constant value. That constant just happens to be 1189 here.

Let any four terms of such a sequence be a, b, $a + b$, $a + 2b$.

Then it is easy to show that $a^2 + ab - b^2 = \pm 1$.

By choosing the order in which the products are subtracted, each result can be made positive. For the Fibonacci sequence, this constant is 1, for the modified sequence here it is 1189. It is important to realize that one may not always transform a portion of one Fibonacci type sequence into a portion of another by reversing digits. It is necessary that in the addition required to build the original sequence, no carrying is required. For example, 11, 21, 32, 53, 85 transforms nicely into 11, 21, 23, 35, 58 with "constants" of 89 and 109 respectively but transforming 17, 19, 36, 55, 91 into 71, 91, 63, 55, 19 does not successfully yield successive terms of a Fibonacci type sequence.

Sincerely,

Monte J. Zerger

Monte J. Zerger is a professor in the School of Science, Mathematics, and Technology at Adams State College, Alamosa, Colorado.

8.6 Validity of a Fibonacci Identity

For any four consecutive Fibonacci numbers a, b, c, d and a+2b

$a^2 + ab - b^2 = \pm 1$

We can verify the above identity by using the fact that
$b^2 - ac = \pm 1$

where a, b, c are three consecutive Fibonacci numbers.
Doing so we have:

$a^2 + ab - b^2 = |b^2 - ac|$ (since they are equal to ± 1).

$a^2 + ab - b^2 = (c - b)^2 + a(c - a) - b^2$

$= c^2 - 2bc + b^2 + ac - a^2 - b^2$

$= c^2 - 2bc + ac - a^2 = c^2 - 2bc + ac - (c - b)^2$

$= c^2 - 2bc + ac - c^2 + 2bc - b^2 = ac - b^2$

But $ac - b^2 = b^2 - ac$, $b^2 - ac = \pm 1$.

Therefore, $|ac - b^2| = |b^2 - ac| = 1$.

Consequently, $|ac - b^2| = |b^2 - ac| = 1$.

If a and b have odd subscripts, then $a^2 + ab - b^2 = 1$.

If a and c have even subscripts, then $a^2 + ab - b^2 = -1$.

hence $|a^2 + ab - b^2| = 1$.

8.7 The Amazing Difference

(Special Properties of $u_7, u_8, u_9, u_{10}, u_{11}$)

Let u_n represent the Fibonacci sequence.
With this notation then, $u_7 = 13$, $u_8 = 21$, $u_9 = 34$, $u_{10} = 55$, $u_{11} = 89$.
Now let us exchange the digits of $u_7 = 13$, $u_8 = 21$, $u_9 = 34$, $u_{10} = 55$, $u_{11} = 89$.
Doing so results to an entirely new sequence.

Call it k_n. Therefore, $k_n = \{31, 12, 43, 55, 98\}$.

For the new sequence, the following are true:

$(31 \times 55) - (12 \times 43) = 1705 - 516 = 1189$
$(43 \times 98) - (55 \times 55) = 4214 - 3025 = 1189$
$(43 \times 55) - (12 \times 98) = 2365 - 1176 = 1189$
$(43 \times 43) - (12 \times 55) = 1849 - 660 = 1189$
$(31 \times 43) - (12 \times 12) = 1333 - 144 = 1189$

8.8 A Search for Counter Examples

(a) Can you identify any four consecutive Fibonacci numbers

$u_n, u_{n+1}, u_{n+2}, u_{n+3}$

such that $\frac{u_{n+3}}{2} + \frac{u_n}{2} \neq u_{n+2}$?

(b) Can you identify any four consecutive Fibonacci numbers

$u_n, u_{n+1}, u_{n+2}, u_{n+3}$

such that $\frac{u_{n+3}}{2} - \frac{u_n}{2} \neq u_{n+1}$?

Chapter 9

Out of Curiosity: Some Beautiful Number Patterns

9.1 Objectives

At the end of the lesson, the students should be able to:

(a) explore and discover more number patterns
(b) express triangular numbers in a general form
(c) explore patterns with triangular numbers
(d) explore patterns with Lucas numbers
(e) explore patterns with Fibonacci numbers

9.2 Introduction

This chapter includes the following:

- Exploring patterns with triangular numbers
- Expressing the nth triangular number
- Deriving the nth triangular number using partial sums of first n natural numbers
- Expressing triangular numbers as partial sums
- Exploring patterns with Lucas numbers
- Exploring patterns with Fibonacci numbers

Let us get started by considering the product of a set of natural numbers and the set of the numbers {2, 3, 4, 5, 6, 7, 8, 9, . . . }

$1(2) = 1(1+1)$	$6(7) = 6(1+6)$
$2(3) = 2(1+2)$	$7(8) = 7(1+7)$
$3(4) = 3(1+3)$	$8(9) = 8(1+8)$
$4(5) = 4(1+4)$	$9(10) = 9(1+9)$
$5(6) = 5(1+5)$	$10(11) = 10(1+10)$

The above in turn can be expressed differently as products of natural numbers and sum of two numbers. Each of the products can be written as n(1+n).

This gives us the general form of the nth triangular number which is given by $\frac{n(n+1)}{2}$.

Exploring Patterns with Triangular Numbers

Following in the next section are other forms of expressing the nth triangular number.
Let us express each member of the set of triangular numbers as a product of two numbers such that one of the numbers is a natural number and the other is an odd positive integer.

$1 = 1\times1 = 1(1\times2) - 1)$	$36 = 4\times9 = 4(4\times2) + 1)$
$3 = 1\times3 = 1(1\times2) + 1)$	$45 = 5\times9 = 5(5\times2) - 1)$
$6 = 2\times3 = 2(2\times2) - 1)$	$55 = 5\times11 = 5(5\times2) + 1)$
$10 = 2\times5 = 2(2\times2) + 1)$	$66 = 6\times11 = 6(6\times2) - 1)$
$15 = 3\times5 = 3(3\times2) - 1)$	$78 = 6\times13 = 6(6\times2) + 1)$
$21 = 3\times7 = 3(3\times2) + 1)$	$91 = 7\times13 = 7(7\times2) - 1)$
$28 = 4\times7 = 4(4\times2) - 1)$	$105 = 7\times15 = 7(7\times2) + 1)$

The above are triangular numbers of the form: $n(2n\pm1)$

(a) n(2n - 1) are triangular numbers of odd subscripts
(b) n(2n + 1) are triangular numbers of even subscripts

Expressing the nth Triangular Number

Triangular numbers of odd subscripts can be expressed as:

$(1+n)\{(2 \times n) +1\}= (1+n)2n+1)$

$= (n+1)(2n+1) = 2n^2 + 3n + 1, n \leq 0$

Similarly, triangular numbers of even subscripts can be expressed as:
$(n - 1)(2n - 1) = 2n^2 - 3n + 1, n \geq 2$.

n	**$2n^2 + 3n + 1$**	**Odd Subscript**
0	1	1
1	6	3
2	15	5
3	28	7
4	45	9
5	66	11
6	91	13
7	120	15
8	153	17
n	**$2n^2 - 3n + 1$**	**Even Subscript**
2	3	2
3	10	4
4	21	6
5	36	8
6	55	10
7	78	12
8	105	14
9	136	16
10	171	18

Table 9.1: Triangular numbers with odd and even subscripts

n	First n Lucas Numbers (a_n)	Sum of Two Alternate Fibonacci Numbers Pattern
1	1	0 + 1
2	3	1 +2
3	4	1 + 3
4	7	2 + 5
5	11	3 +8
6	18	5 + 13
7	29	8 + 21
8	47	13 + 34

Table 9.2: Renaming first n Lucas numbers as sum of two alternate Fibonacci numbers

In this table, how does a particular a_n relate to a particular u_n?

n	Sum of Squares of first *n* triangular numbers	Evaluated Sum (Triangular Number)	Pattern (Subscript)
1	$1^2 + 3^2$	10	4
2	$3^2 + 6^2$	45	9
3	$6^2 + 10^2$	136	16
4	$10^2 + 15^2$	325	25
5	$15^2 + 21^2$	666	36
6	$21^2 + 28^2$	1225	49

Table 9.3: Sum of squares of two consecutive triangular numbers

n	Difference of Squares of first *n* triangular numbers	Evaluated Difference	Pattern
1	$3^2 - 1^2$	8	4
2	$6^2 - 3^2$	27	9
3	$10^2 - 6^2$	64	16
4	$15^2 - 10^2$	125	25
5	$21^2 - 15^2$	216	36
6	$28^2 - 21^2$	512	49

Table 9.4: Difference of squares of two consecutive triangular numbers

EXAMPLE

The difference between two consecutive triangular numbers is 15. Find the numbers.

SOLUTION

Let the two consecutive triangular numbers be y and x ($x > y$).

Therefore, $x - y = 15$ ....(i).

Also, for any two consecutive triangular numbers y and x,
$x + y = (y-x)^2 = y^2 - 2yx + x^2$ ....(ii)

From equation (i), $(x-y)^2 = 15^2$
$x^2 - 2xy + y^2 = 225$ ....(iii)

By subtracting equation (iii) from equation (ii) we have:

$y^2 - 2yx + x^2$
$\underline{225 = x^2 - 2xy + y^2}$
$x + y = 225 = 0 \Leftrightarrow x+y = 225$ ....(iv)

Combining equations (i) and (iv) we have:

$x - y = 15$ ....(i)
$x+y = 225$ ....(iv)

By adding equations (i) and (iv) we have:

$2x = 240$. From here $x = 120$.

From equation (i), $120 - y = 15$.

n	a_n	b_n	b_n-a_n
1	3	3	2
2	3	6	3
3	6	10	4
4	10	15	5
5	15	21	6
6	21	28	7
...	...	...	...
...	...	...	...

Table 9.5: Difference between two consecutive triangular numbers

Study the pattern in table 9.5 and complete any missing information. In words, how do you describe the relation between the first n triangular numbers and their differences?

Find $\sqrt{a_n - b_n}$ for all values of n.

What do you notice?

From here, deduce that $a_n + b_n = (b_n - a_n)^2$.

n	Partial Sums of First n Even Numbers Whose First Term is 2
1	2 =2
2	2 + 4= 6
3	2 + 4 + 6 = 12
4	2 + 4 + 6 + 8 = 20
5	2 + 4 + 6 + 8 +10 = 30
6	2 + 4 + 6 + 8 + 10 +12 = 42
7	2 + 4 + 6 + 8 + 10 +12 + 14 = 56

Table 9.6: Partial sums of first n even numbers whose first term is 2

When the above results are divided by 2, the result is a set of triangular numbers.

Again! Yes, again. Wow!

The above situation can be generally expressed as

$$\frac{n(n-1)}{2}, n^3 \; 2$$

By mathematical induction, the nth triangular number is given by:

$$\frac{n(+1)(n-1+1)}{2} = \frac{n(n+1)}{2}$$

Partial Sums of First n Positive Even Numbers Whose First Term is 2

$2=2=1^2+1$
$2+4 = 6 = 2^2 + 2$
$2+4+6 =12 = 3^2 + 3$
$2+4+6+8 =20 = 4^2 +4$
$2+4+6+8+10 = 30 = 5^2 +5$
$2+4+6+8+10+12= 42 = 6^2+6$
$2+4+6+8+10+12+14 = 56 =7^2 +7$
$2+4+6+8+10+12+14+16 = 72 = 8^2+8$

Each of the above can generally be written as $n^2 + n$ and when divided by 2 we have:

$$\frac{n^2+n}{2} = \frac{n(n+1)}{2}$$

n	Partial Sums of First n Natural Numbers (w_n)
1	1=1
2	1+2=3
3	1+2+3=6
4	1+2+3+4=10
5	1+2+3+4+5=15
6	1+2+3+4+5+6=21
7	1+2+3+4+5+6+7=28

Table 9.7: Partial sums of first n natural numbers

Subset of the set of Natural Numbers	*n*	*A*	*L*+1	*a*+*L*	*n*+1
1	1	1	2	2	3
1, 2	2	1	3	3	3
1, 2, 3	3	1	4	4	4
1, 2, 3, 4	4	1	4	4	4
1, 2, 3, 4, 5	5	1	6	6	6
1, 2, 3, 4, 5, 6	6	1	7	7	7
1, 2, 3,4,5, 6, 7	7	1	8	8	8

Table 9.8: Demonstrating that L+1 = a + L.

The sum of each subset is given by

$$\frac{n(n+1)}{2}$$

But n+1 = a + L.

By substitution, the sum of each subset is given by $\frac{n(n+1)}{2}$ and this is the general form of expressing the nth triangular number. Out of curiosity, let us find the sum of 1 and twice individual triangular numbers. Take a look

Addition Facts	Addition Facts Renamed
4	$4 = 2^2 + 0 = (1+1)^2$
4+6	$10 = 3^2 + 1 = (2+1)^2 + 1$
4+6+8	$18 = 4^2 + 2 = (3+1)^2 + 2$
4+6+8+10	$28 = 5^2 + 3 = (4+1)^2 + 3$
4+6+8+10+12	$40 = 6^2 + 4 = (5+1)^2 + 4$
4+6+8+10+12+14	$54 = 7^2 + 5 = (6+1)^2 + 5$
4+6+8+10+12+16	$70 = 8^2 + 6 = (7+1)^2 + 6$

Table 9.9: Addition Facts Re-named.

This addition resulted to the set of numbers {1, 3, 7, 13, 21, 31, 43, ⋯}

The set {1, 3, 7, 13, 21, 31, 43, ⋯} is itself a pattern though it doesn't seem to be one. What is our objective here? Since {1, 3, 7, 13, 21, 31, 43, ⋯} is a sequence, let us investigate and derive the nth term of the sequence.

Addition Facts	Addition Facts Renamed	Addition Facts Renamed	Addition Facts Renamed
3	3+0	$3 + 1^2 - 1$	$3 + (0 + 1)^2 - 1$
3+4	3+4	$3 + 2^2 + 0$	$3 + (1 + 1)^2 + 0$
3+4+6	3+10	$3 + 3^2 + 1$	$3 + (2 + 1)^2 + 1$
3+4+6+8	3+18	$3 + 4^2 + 2$	$3 + (3 + 1)^2 + 2$
3+4+6+8+10	3+28	$3 + 5^2 + 3$	$3 + (4 + 1)^2 + 3$
3+4+6+8+10+12	3+40	$3 + 6^2 + 4$	$3 + (5 + 1)^2 + 4$
3+4+6+8+10+12+14	3+54	$3 + 7^2 + 5$	$3 + (6 + 1)^2 + 5$
3+4+6+8+10+12+14+16	3+70	$3 + 8^2 + 6$	$3 + (7 + 1)^2 + 6$

Table 9.10: Addition Facts Re-named.

Using your newly derived formula, find the 100th term of the sequence. Could we have found the nth term differently?
The answer is yes.
The search for the nth term of the sequence continues.

Let us use a few of the above addition facts to further our search for the nth term of the sequence. Let each addition fact be a case and we are going to consider only four such cases:

Case 1: 3+4+6+8+10
Case 2: 3+4+6+8+10+12
Case 3: 3+4+6+8+10+12+14
Case 4: 3+4+6+8+10+12+14+16

Case 1

The number of terms in (3+4+6+8+10) = 5
But $3+(4+6+8+10) = 3+(4+1)^2+3$
$4 + 1 = 5, 5 - 2 = 3.$

If 5 = n, then 4+1 = (n-1) + 1 and n-2 = 3.

Since $3+4+6+8+10 =3 + (4 + 1)^2 + 3$, by substituting in $3 + (4 + 1)^2 + 3$, we have:
$3 + (4 + 1) + 3 = 3 + [(n - 1) + 1]^2 + (n - 2)$
$=3 + n^2 + n - 2 = n^2 + n - 2 + 3 = n^2 + n + 1$

The same holds water for the rest of the cases in table 9.10.

Therefore, $n^2 + n + 1$ is a generalization for the nth term of the sequence, 1, 3, 7, 13, 21, 31, 43 ⋯
where $n \geq 0$.

Case 2

The number of terms in (3+4+6+8+10+12) = 6.
But $3+(4+6+8+10+12) = 3+(5+1)^2+4$
$5 + 1 = 6, 6 - 2 = 4.$

If 6 = n, then 5+1 = (n-1) + 1 and n-2 = 4.

Since $3+4+6+8+10+12 =3 + (5 + 1)^2 + 4$, by substituting in $3 + (5 + 1)^2 + 4$, we have:
$3 + (5 + 1) + 3=3 + [(n - 1) + 1]^2 + (n - 2)$
$=3 + n^2 + n - 2 = n^2 + n - 2 + 3 = n^2 + n + 1$

The same holds water for the rest of the cases in table 9.10.
Therefore, $n^2 + n + 1$ is a generalization for the nth term of the sequence, 1, 3, 7, 13, 21, 31, 43 ⋯
where $n \geq 0$.

Case 3

The number of terms in (3+4+6+8+10+12+14) = 7.
But $3+(4+6+8+10+12+14) = 3+(6+1)^2+5$.
6 + 1 = 7, 7 - 2 = 5.

If 7 = n, then 6+1 = (n-1) + 1 and n-2 = 5.

Since $3+4+6+8+10+12+14 =3 + (6 + 1)^2 + 5$, by substituting in $3 + (6 + 1)^2 + 5$, we have:
$3 + (6 + 1) + 3=3 + [(n - 1) + 1]^2 + (n - 2)$
$=3 + n^2 + n - 2 = n^2 + n - 2 + 3 = n^2 + n + 1$

The same holds water for the rest of the cases in table 9.10.
Therefore, $n^2 + n + 1$ is a generalization for the nth term of the sequence, 1, 3, 7, 13, 21, 31, 43 ⋯
where $n \geq 0$.

Case 4

The number of terms in (3+4+6+8+10+12+14+16) = 8
But $3+(4+6+8+10+12+14+16) = 3 + (7 + 1)^2 + 4$
7 + 1 = 8, 8 - 2 = 6.

If 8 = n, then 7+1 = (n-1) + 1 and n-2 = 6.

Since $3+4+6+8+10+12 =3+(7+1)^2 + 4$, by substituting in $3+(7+1)^2 + 6$, we have:
$3+(7+1)^2+6=3+ [(n - 1) + 1]^2 + (n - 2)$
$= 3 + n^2 + n - 2 = n^2 + n - 2 + 3 = n^2 + n + 1$

The same holds water for the rest of the cases in table 9.10.

Therefore, $n^2 + n + 1$ is a generalization for the nth term of the sequence, 1, 3, 7, 13, 21, 31, 43 ⋯
where $n \geq 0$.

n	Sum of squares of first n triangular numbers	Evaluated Sum	Pattern
1	1^2+3^2	10	4
2	3^2+6^2	45	9
3	6^2+10^2	136	16
4	10^2+15^2	325	25
5	15^2+21^2	666	36
6	...	...	...

Table 9.11: Sum of squares of first n Fibonacci numbers

In table 9.11, state in your own words how each evaluated sum is related to a particular triangular number. The use of calculator is highly encouraged.

n	Sum of First n Lucas Numbers	Evaluated Sum	Pattern
1	1	1	4-3=1
2	1+3	4	7-3=4
3	1+3+4	8	11-3=8
4	1+3+4+7	15	18-3=15
5	1+3+4+7+11	26	29-3=26
6	1+3+4+7+11+18	44	47-3=44
7	1+3+4+7+11+18+29	73	76-3=73
8	1+3+4+7+11+18+29+47	120	123-3=120

Table 9.12: Partial Sums of first n Lucas numbers

(a) Lucas numbers are numbers of the form: 1, 3, 4, 7, 11, 18, 29, 47, 78···

Study table 3 to table 9.12.
Use the pattern to answer these questions.
State the general rule for finding the sum of first n Lucas numbers.

(b) State in symbols how to find the sum of the first 100 Lucas numbers.
6, 36, 120, 300, 630, 1176, ··· are triangular numbers with 3, 8, 15, 24, 35, 48 ··· as subscripts.

n	Triangular Number	Subscript	Subscript as a product
1	6	3	1×3
2	36	8	2×4
3	120	15	3×5
4	300	24	4×6
5	630	35	5×7
6	1176	48	6×8

Table 9.13: Triangular numbers with subscripts 1 less than a square number

Generally, the subscripts of the above triangular numbers can be stated in three ways:

(a) (b) (c)

Prove that all the three forms express a particular subscript.

n	Sum of Two Consecutive Terms of g_n	Evaluated Sum	Pattern
1	0+3	3	1×3
2	3+5	8	2×4
3	5+16	21	3×7
4	16+39	55	5×11
5	39+105	144	8×18
6	105+272	377	13×29
7	. . .	. . .	. . .

Table 9.14: Sum of two consecutive Fibonacci numbers as product of u_n and a_{n+1}

n	Differences of Squares of First n Fibonacci Numbers	Evaluated Difference	Pattern
1	1^2-1^2	0	0×2
2	2^2-1^2	3	1×3

3	3^2-2^2	5	1×5
4	5^2-3^2	16	2×8
5	8^2-5^2	39	3×13
6	13^2-8^2	105	5×21
7	...	...	...
8	...	...	...

Table 9.15: Difference of squares of first n Fibonacci numbers

Study the above table. In words, describe how each evaluated difference is related to a particular Fibonacci number.

9.3 A Search for Counter Examples

(a) Can you identify any triangular number of even subscript that cannot be generated by the relationship, $2n^2 - 3n + 1, n \geq 2$?

(b) Can you identify any triangular number of odd subscripts that cannot be generated by the relationship, $2n^2 + 3n + 1, n \geq 0$?

(c) Can you identify any triangular number that cannot generally be represented by $\frac{n(n+1)}{2}$?

(d) Can you identify any partial sum of first *n* triangular numbers whose sum is not equal to a triangular number whose subscript is equal to *n*?

(e) Can you identify any two consecutive triangular numbers *a* and *b* whose difference of squares is not equal to n^3?

(f) Can you identify any even-subscripted triangular number that cannot be represented generally as $n(2n-1)$?

(g) Can you identify any two consecutive triangular numbers *a* and *b* with subscripts *n* - 1 and n respectively where $(b - a)^3$ is not equal to n^3?

Chapter 10

Mathematics Behind Bars: My Experience with the US Immigration

10.1 Objectives

At the end of the lesson, the students should be able to:

(a) find the roots of the equation $n^2 - n - k = 0$.
(b) find explicitly, the sum and product of the roots of the equation, $n^2 - n - k = 0$.
(c) find implicitly, the sum and product of the roots of the equation $n^2 - n - k = 0$.
(d) verify the roots of the equation, $n^2 - n - k = 0$.
(e) share and learn from my personal experience with the United States Immigration

10.2 Introduction

This chapter is based on the actual personal experience of the author himself. It is a story about his confinement at Oklahoma City County Jail by the United States Immigration & Naturalization Service (INS). For purpose of privacy, Diana and Oprah are fictional characters. They were used to inject some sense of fiction into the story. The town of Potomico does not exist. New Jevico does not exist either. The water treatment plant exists only in the imagination of the author. Any resemblance in real life therefore, is unintentional.

For every adversity, there is a seed of an equivalent benefit. These are the words of Napoleon Hill, the author of "*Think and Grow Rich*", a book that has revolutionized many lives all over the world. I never realized the reality of the above statement by Napoleon Hill not until April 10, 1987. What actually happened?

10.3 Apprehension by the United States Immigration

On April 10, 1987, I was taken into custody by The United States Immigration & Naturalization Service (INS). Why? Why was I apprehended by the INS?

Drug trafficking? No!
Forgery? No!
Insurance scam? No!
None of those. Hold your patience. I will tell you.
Before then, one more question needs to be answered.

10.4 Coming to the United States

First of all, my goal here is not to portray the US Immigration negatively. My goal is to tell my story just like anyone else who has a story to tell. I came to the United States in the fall of 1984 as a Nigerian immigrant on a non-immigrant visa (a more honorable way of avoiding the use of "visiting visa" which most people in the Nigerian or African community in the US immediately equate with deportation) with the sole objective to explore opportunities to publish my manuscript. (35,000 Nigerians applied for non-immigrant visa between October 1, 1989 and September 20, 1990). When the same question came up at No. 2 Eleke Crescent, Victoria Island, Lagos, (the home of the US Embassy in Nigeria), with the embassy officials, my answer was not different. Despite my genuine reasons for petitioning for an entry visa into the United States, my request was denied on July 9, 1984 for what was described as "incomplete documentation." However, my visa was finally granted on August 24, 1984 with an expiration date of December 14, 1984. This means that I had only 136 days to stay on arrival. My first port of call was JFK

International Airport, New York from where I headed for Minneapolis, St. Paul, Minnesota via LaGuarda Airport.

10.5 Legalizing My Stay in the United States

In the process of legalizing my stay (my visa having expired at this time), I was apprehended by one of the most feared immigration authorities in the world, The United States Immigration & Naturalization Service. From henceforth, I was declared an illegal alien, booked into Oklahoma City County Jail and held on a bond in the amount of $2,500. Illegal, not in the sense that I swam from across the border. Illegal in the sense that my formal papers had expired.

Figure 10.1: Map Showing Oklahoma City County Jail

The bond was eventually reduced to $1,000 through the untiring efforts of my attorney, Mr. Michael Smith of Oklahoma City. The only pilgrimage I was taught in school and therefore, know about is that of Mecca (by Moslems) or Bethlehem (by Christians) but to some people, OCJ (acronym for Oklahoma City County Jail) is another pilgrimage which they have to undertake at least once a year.

10.6 US Immigration Visits My Home in Nigeria

While I was in jail, the US immigration, in co-operation with the US Embassy in Lagos, Nigeria, visited my home in Nigeria twice in May of 1987. They talked to my parents and some relatives. They also paid visit to my Alma Mater, Alvan Ikoku College of Education, Owerri. The purpose of the visit was to verify my criminal records. This was to justify, quicken and complete my deportation proceedings. They also verified my marital status. There were no criminal records and there were no marital irregularities because I did not have a wife in Nigeria at the time of the investigation.

10.7 Diana Goes on Vacation

Diana has been running a busy schedule at the office. She is a secretary with a water treatment company downtown Potomico. She was happy about her vacation plans because her time, quite unlike in the office, will be free from those incessant phone calls, running of errands, handling of files and other monotonous activities which normally punctuate her life at the office. Those who have had the opportunity of working in such an office know what we are talking about here. Diana was therefore, jubilantly, looking forward to her new experience. She was just receiving her first pay check as a secretary, having recently left her former job.

The same time last year, Diana spent her vacation at Jevico Hotels in New Jevico. She had again decided to go to Jevico, despite the fact that her friend, Oprah, wanted to talk her out of it. (Oprah preferred Hawaii and Alaska) to New Jevico. Oprah had spent her vacation at Hawaii for three consecutive years now.

One evening, as Diana was stepping into her hotel room, she was interrupted by a team of plain clothes detectives, who were then investigating a case of tax evasion by a well-known business woman. Diana and three other tenants, were bundled into a waiting van for further interrogation at New Jevico police station. A day's detention spilled into twenty days. Little did Diana know that she would not be back to her hotel room the same day.

10.8 Positive Mental Attitude is the Key

That day was April 10, 1987. One of the new arrivals, (a black teenager, about sixteen years of age) Michelle, (not his real name) went straight to the calendar which was centrally hung on the tall structure of a bunker bed bolted to the decking of the multipurpose building for rigid support.

"One, two, three, four, five, six, seven, and eight . . . from today will be the 7th and I will be out of here," she said. This was after her arraignment.

None of the inmates took her words seriously for the day he was booked in, he told others that he would be released after three days advising others to exhibit some positive mental attitude toward their respective circumstances. He said this as soon as he made one of the two phone calls which each inmate is allowed as soon as he or she is checked into the facility. Mitchelle was drawing some inspiration from the Bible of how Christ rebuilt the temple in three days. Now, three days had come and gone. His release never came.

10.9 To Whom Much is Given, Much Is Expected

There was an inmate who deserves mention here. His name was Sheti. He was popular among some members of the detention staff. To whom much is given, much is expected. "I never experienced the reality of this statement of my teacher the second time except when I met this young man," Diana continued. He always received extra food in recognition for his extra mile of service—sweeping and moping the floor first thing in the mornings (pacing the floor up and down, talking to herself).

"How long have you been here?", an inmate asked another inmate named Barbara.
"Oh, about four weeks," Barbara said.
"When are you leaving this place?" the inmate continued.
"Going out of here?"
Barbara repeated as if she was not within hearing distance.
"Yes," the inmate said.
"You dare not say that."

I will be transferred from here to Lexington Penitentiary where I . . ."
"You are not going yet?" (the inmate cuts in).
"No," Barbara replied.

At this point, she did not want to encourage the story telling even though Barbara had more terrifying stories to tell. Many stories are told in detention camps but they never see the light of day.

10.10 Life in a Detention Camp

Life in a detention camp is not something one can be proud of. Monotony is enough to make it a hell on earth. To break the monotony, inmates occasionally engage themselves in all sorts of activities including press-ups, shadow boxing, playing cards, and watching television. They alternate between these activities, all geared toward breaking the continued boredom. To get out of this rut, Diana decided to feast her eyes on traffic along Hudson Street and Walker Avenue, downtown Oklahoma City, about two blocks from Alfred P. Murray Federal building that was gutted by a terrorist bomb on April 19, 1995, leaving 168 dead and several hundreds injured.

Vertical and horizontal bars naturally created spaces in between and these bars served as two of the four walls of the detention room. Through these spaces, one could catch a glimpse of what goes on in some parts of these streets visible from the eleven story building. It is one of the outstanding buildings in the area and some people pay homage to it at least once a year as if it was another Mecca or Jerusalem, the holy cities.

10.11 The Eureka Moment

Looking out through these spaces in a desperate effort to while away time and expend some pent-up energy, something attracted her attention. The object of attention was the ventilation bars on which she was leaning for support. "There might be something interesting with these bars," she had once told herself. With continued observation, she decided to count the number of bars within a specific length.

"Wait a minute," she shouted as in Archimedes' eureka moment.

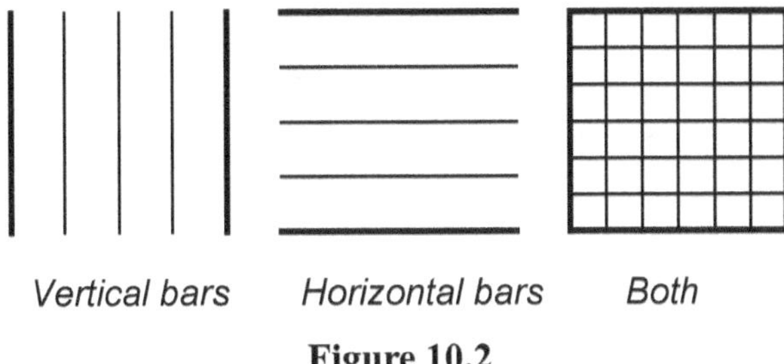

Figure 10.2

10.12 Don't Give Up: Giving Up is Not an Option

"Adversity, though not an experience we eagerly look forward to, might carry with it some ideas that will likely help us in solving a problem or completing a project," Diana said. Don't give up, even though giving up is an option. Is it the best option? The answer is no. Has it ever been the best option? The answer is also no. Will it ever be the best option? Of course, the answer is still no. Diana is inviting us to share the following experience with her and hopes you will always strive to use your time gainfully, no matter where you are and the prevailing circumstances.

10.13 The Forgotten Difference

Let us go back to the bars that caught Diana's attention.

She was noticing that some consistency was developing. She repeated the count many times, each time noting the number of intervals between them. Each time she counted them, the number of intervals between the bars was always one less than the number of bars. At this point, Diana decided to make further investigation, taking note of a "valuable" idea on paper. Having been provided with no paper and pencil or pen with which to play with, she borrowed used envelopes and scrap paper from fellow inmates. She also found some court papers in her possession very useful by writing on any available white space.

Number of Poles *(n)*	Number of Intervals *(n-1)*	Difference between *n* and *n*-1 *n-(n-1)*
1	0	1
2	1	1
3	2	1
4	3	1
5	4	1

Table 10.1: Difference between *n* and *n - 1*

With a series of observations, *n* represent the number of electric poles. As she continued the investigation, she noted that the difference between the number of poles and the number of intervals is always equal to 1. As the investigation progressed, her interest increased as did the urge to gain insight into any subtle mathematical ideas. The "pole problem" as she chose to describe this situation, could be stated as follows:

Given n poles spaced out in a straight line either vertically or horizontally at equal intervals, there are n - 1 intervals between them and the product of the number of poles and the number of intervals is always equal to an even number.

This investigation eventually led to such conventional classroom topics as quadratic equations, perfect squares, properties of positive odd and even integers, negative and positive roots of a quadratic equation, arithmetic series, etc. The list is not exhaustive. More can be added.

10.14 Analyzing the Product, $n(n-1)$

Throughout this chapter, n stands for the number of poles and n-1 is therefore, the number of intervals between the poles. The product of the two is represented by k so that $n(n-1) = k$ and is always an even number.

1. Can this be proved? The answer is yes.

2. Having seen that $n(n-1) = k$, can we derive a formula in terms of k by which n can be found with any given value of k?

That becomes our second question.
Now why is $n(n-1)$ always an even number?
To answer this question, we have two cases to consider.

Proof by Cases

Case 1: When n is Odd

What does this mean?
It means that n-1 contains a factor of 2 so that the product of n and $n-1$ will yield an even number. An even number, it should be recalled, is always expressed as a factor of 2 and some number p where p is an integer.

Consequently, $n(n-1) = 2p - 1(2p) = 4p^2-2p = 2p(2p-1)$.

Therefore, $n(n-1)$ is an even number.
When n is odd, $n-1$ must be even

Case 2: When n is even

When n is even, $n-1$ must be odd. It follows that n must be a multiple of 2. Therefore, n can be written as some number $2p$. Therefore, $n(n-1) = 2p(2p-1) = 4p - 2p$ and $4p - 2p$ is even since it can be written as $2p(2p - 1)$ which is an even number since $2p(2p-1)$ has 2 as a factor. Therefore, $n(n-1)$ is an even number.

Therefore, the product of the number of intervals and the number of poles is always an even number. The second question asks:

Can we derive a formula in terms of k by which n can be found with any given value of k?

Now let us look at the equation, $n(n\text{-}1) = k$.
We can re-write n(n-1) as $n^2 - n - k = 0$.
This equation is a quantity involving a power of 2.
It is therefore, a quadratic equation.

If $n^2 - n - k = 0$ is a quadratic equation, how do we solve for n?
$ax^2 + bx + c$ and $n^2 - n - k = 0$

What do $ax^2 + bx + c$ and $n^2 - n - k = 0$ have in common?
Generally, a quadratic equation is of the form $ax^2+bx + c = 0$.
Now since $n^2 - n - k = 0$ is a quadratic equation, the following identity is true:

$n^2 - n - k \equiv ax^2 + bx + c.$

Do you agree?
There are similarities between $n^2 - n - k$ and $ax + bx + c$.

What is it?
They are identical. If $n^2 - n - k \equiv ax^2 + bx + c$, the following are true statements:
$n = ax^2$, *(b)* $- n = bx$ *(c)* $c = - k$

Generally given the equation, $ax^2 + bx + c = 0$,

$$x = \frac{-b \pm \sqrt{b^2 - 4ac}}{2a}$$

This is the general solution of the equation of the form, $ax^2 + bx + c = 0$.
Now look at $n^2 - n - k = 0$. What do you notice?

10.15 Comparing Coefficients and Constant Terms

(a) If $n^2 = ax^2$, it implies that $a = 1$ since the coefficient of n^2 is equal to 1.

(b) If $-n = bx$, it implies that $b = -1$ since the coefficient of n is equal to the coefficient of x.

(c) If $-k = c$, then the value of c in $n^2 - n - k$ is $-k$.

(The corresponding constant terms are equal)

10.16 Solving the Equation, $n^2 - n - k = 0$

Now we can apply the general solution formula having known the values of a, b, c in $n^2 - n - k = 0$ through a process of comparing coefficients and constants.
By substitution, we have:

$$x = \frac{1 \pm \sqrt{(-1)^2 - 4.1.(-k)}}{2}$$

$$= \frac{1 \pm \sqrt{1 + 4k}}{2}$$

$$\frac{1 \pm \sqrt{4k + 1}}{2}$$

Therefore, $n = \dfrac{1 \pm \sqrt{4k+1}}{2}$.

It should be noted that n has two values. Find them.

Which one of these satisfies our value for n with respect to the pole problem?

Using the above formula, find n when $k = 6$, 20, and 42.

Can you find another expression in terms of k which could be used to find the values of n?

10.17 Verifying the Roots of the Equation, $n^2 - n - k = 0$

Objectives

To verify that $\frac{k}{n-1}$ and $\frac{-k}{n}$ are the roots of the equation, $n^2 - n - k = 0$.

Revise the general form of a quadratic equation, $ax^2 + bx + c = 0$.

Compare the general form with $n^2 - n - k = 0$.

10.18 Comparing $n^2 - n - k = 0$ with $ax^2 + bx + c = 0$

If $ax^2 + bx + c = 0$, the following are true:

(a) Sum of roots = $\frac{-b}{a}$

(b) Product of roots = $\frac{c}{a}$

If $\frac{k}{n-1}$ and $\frac{-k}{n}$ are the roots of the equation, $n^2 - n - k = 0$,

can we relate it to the case of $ax^2 + bx + c = 0$?

Of course, the answer is yes, we can.

Now the answer is yes, what's next?

10.18 Finding Sum and Product of Roots Implicitly

Now the answer is yes, we have to identify a, b, c in $n^2 - n - k = 0$.

Doing so we have: $a = 1$, $b = -1$, $c = -k$

We can now also find the sum and product of roots with $\frac{-b}{a}$ as sum of roots and $\frac{c}{a}$ product of roots.

Doing so we have:

$\frac{-b}{a} = \frac{-(-1)}{1} = 1$ .... equation (i)

$\frac{c}{a} = \frac{-k}{1} = -k$ .... equation (ii)

10.19 Finding Sum and Product Explicitly

Now taking $\frac{k}{n-1}$ and $\frac{-k}{n}$ as roots of the equation, $n^2 - n - k = 0$,

let us find the product and sum of roots.

Product of roots $= \left(\frac{k}{n-1}\right)\left(\frac{-k}{n}\right) = \frac{n(n-1)}{n} \times \frac{n(n-1)}{n-1}$

$= -(n-1)(n) = -k$ .... **equation (iii)**

Compare our result in equation **(iii)** with our result in equation **(ii)**

Sum of roots $= \left(\frac{k}{n-1}\right) + \left(\frac{-k}{n}\right)$

$= -(n-1) + (n) = 1$ .... equation (iv)

Compare our result in equation (iv) with our result in equation (i).

(a) What does it mean if our result in equation **(ii)** is equal to our result in equation **(iii)**?

(b) What does it mean if our result in equation **(i)** is equal to our result in equation **(iv)**?

10.21 Chapter Summary

(a) An even number is a number that is divisible by 2.
(b) An even number can be written as a number $2a$ where a is a whole number.
(c) An odd number is a number that is not divisible by 2.
(d) An odd number can be represented as $2a+1$ where a is a whole number or $2a - 1$ where a is a natural number.

The product of an odd number and an even number is always an even number.

If $n^2 - n - k$ is equivalent to $ax^2 + bx + c$, then the following are true statements:

$n^2 \equiv ax^2, n \equiv bx, c \equiv k.$
If $n^2 - n - k = 0$, then

$$n = \frac{1+\sqrt{4k+1}}{2}$$

10.22 Oral Exercises

Answer True or False to the following questions

(a) The square of a positive odd number is also an odd number.
(b) All real numbers are whole numbers
(c) All whole numbers are real numbers
(d) All even numbers are whole numbers
(e) All whole numbers are also even numbers
(f) A positive odd integer can be expressed in the form 2a+1 where a is a natural number.

10.23 Written Exercises

1. Answer True or False to the following questions.

   (a) If n is even, then n(n-1) must be odd.
   (b) If n is either odd or even, then $n - 1$ is even.
   (c) n(n-1) is a quadratic equation.
   (d) $n(n-1) = 0$ has two solutions
   (e) If n is even, then $n(n+1)$ is also even.
   (f) If n is odd, then n - 1 must be odd.
   (g) The sum of an odd number and an even number is always even.

2. Prove that the difference between any two consecutive square numbers is an odd number.

3. Find the successive ratios of n and $\frac{n}{n-2}$ as n becomes large? How can we express this mathematically?

4. For what values of n is $\frac{1}{2}$(number of poles)=(number of intervals) a true statement?

5. A quadratic equation is generally expressed as $ax^2 + bx + c = 0, a \neq 0$.

   (a) What is the condition that $n^2 - n - k = 0$ always has two unequal real roots?
   (b) What is the value of the discriminant, $b^2 - 4ac$ with respect to $n^2 - n - k = 0$?

6. In the expression, $n = \frac{1 \pm \sqrt{4k+1}}{2}$,

   (a) Find an expression for k in terms of a number c.
   (b) What is the minimum value of c satisfying the above?
   (c) What value of k satisfies the equation, $b^2 - 4ac = 0$ given that $n^2 - n - k = 0$?

7. What is the restriction to be imposed on b if $2b + 1$ is to represent an odd number?

   (a) If $r = \frac{n}{n-1}$, find n in terms of r.

   (b) When is k a perfect square? (with respect to circle problem)

   (d) Why is a circle of interest with regard to n and $n - 1$ as used in this chapter?

   (e) In a circle, find the numerical value of c, given that r = $r = \frac{n}{n-1}$.

   (f) Is the formula, $n = \frac{1 \pm \sqrt{4k+1}}{2}$ true for circles? If not why?

8. (a) What are the values of n when $k = 0$? (b) Can $b^2 - 4ac$ ever be equal to zero in the context of the pole problem?

10.24 A Search for Counter Examples

(a) Can you identify any quadratic equation whose sum of roots and product of roots cannot be found by $\frac{-b}{a}$ and $\frac{c}{a}$ respectively?

(b) Can you identify any quadratic equation that can generally be represented as $ax^2 + bx + c = 0$ where a = 0?

(c) Can you identify any quadratic equation that can be represented generally as $ax^2 + bx + c = 0$ where a ≠ 0?

Chapter 11

Determining Fairness

11.1 Objectives

At the end of the lesson, the students should be able to determine if there is any fairness in the payment options available to customers of a satellite TV call center.

11.2 Introduction

I wrote this article when I was in the employ of Matrixx Marketing in Moore, Oklahoma. Moore is about 7 miles from Norman where I lived at the time. This facility which opened in the fall of 1997 is a call center for DirectTV, a company that among other things, sells satellite TV programming. At the time Matrixx Marketing has two call centers. One was in Norwood, Missouri and the other in Salt Lake City, Utah, the headquarters. That of Norwood was closed after Moore call center was opened. At the Moore facility, salespeople are trained in outbound telemarketing skills that enable them sell DirectTV to customers all across United States. Sports packages as used here refer to different service combinations made available to Matrixx customers.

11.3 Examples of Packages

Package No. 1: Basic Service

Package No. 1: Basic Service + HBO
Package No. 1: Basic Service + Cinemax
Package No. 1: Basic Service + HBO + Cinemax + Showtime
Basic Service comprises of regular TV channels including CNN, MSNBC, C-Span 1 & 2 etc. while HBO, Cinemax and Showtime are just movies.

11.4 Payment Options

These packages, one should expect, have different prices and to pay these bills, customers are provided five payment options. The information on these payment options and other relevant information which are contained in the Help file exist only in 'read only' format, changes cannot be made to these files though they are updated as and when needed by the appropriate company department. If you don't understand the above packages, don't worry about it. Just read on.

11.5 Help Section

This Help section contains update information about Matrixx's Marketing (the promotions within the company etc.) and sales information, new products, special packages etc.

11.6 My Premise

As hours turned into days and days turned into weeks, weeks turned into months, the Help section started to attract my attention in a way different from what the initial purpose was. One day I couldn't handle it anymore. I decided to investigate my thought. What attracted my attention were the various options provided by Matrixx Marketing. I started to look for a pattern.

'The goal of Matrixx Marketing is progressive customer service," our trainer told us in a training session few days before. To me, one of the components of progressive customer service is fairness. Fairness for whom? Fairness for the customers. These five payment options raised a question, a question that needs an answer.

The question is: How fair are payment options? I was not concerned with fairness of individual options. My concern was the fairness of the entire options. To assess the degree of fairness, I came up with a premise, a premise that is embedded in the following fractions:

$$\frac{1}{4}, \frac{2}{4}, \frac{3}{4}, \frac{4}{4}, \frac{5}{4}$$

The above fractions correspond to the five payment options provided to the customers. Looking at an individual payment option, the numerator indicates the number of payments made by a customer while the denominator indicates the number of available options. My premise is that if, when manipulated and subjected to some mathematical operations, and the results present a pattern, (a pattern based not solely just on these fractions, but also on the interpretations associated with them), then we will conclude that the payment options are fair, otherwise, there is no fairness.

You will agree with me that the fractions,

$$\frac{1}{4}, \frac{2}{4}, \frac{3}{4}, \frac{4}{4}, \frac{5}{4}$$

already present a pattern. After all, are these not ascending fractions? Unfortunately, that pattern is not enough. The pattern we have in mind should not necessarily be in form of a sequence or a number series. It is more complex than just a mere sequence. The pattern we have in mind therefore, also goes beyond ordering of fractions based on increasing numerators and common denominators. Before we continue, let us look at the meanings given to the various payment options.

11.7 Giving Meanings to the Five Payment Options

The following options were given these interpretations by the company as follows:

Option No. 1: In this option, the customer made a zero payment. He/she will be billed a month later. This is coded as “1111”.

Option No. 2: The customer agrees to make one payment at a time leaving him/her with three more equal payments. This is coded as "1, 111".

Option No. 3: The customer agrees to make two equal payments at a time leaving him/her with two more equal payments. This is coded as "11,11".

Option No. 4: In this option, the customer agrees to make three equal payments at a time leaving him/her with only one payment. This is coded as "111,1".

Option No. 5: This is the last option. In this option, the customer agrees to make four equal payments at a time. This is what Matrixx Marketing calls "The Confirm Option" and with this the long journey of testing my premise begins.

11.8 Exploring and Testing My Premise

To me these payment options raised a question, a question whose answer will go a long way in determining this degree of fairness we are talking about. The question is: Is there a relationship between an individual option (call it x) and a corresponding number of equal payments (call it y) made by a customer? I decided to investigate and the answer is a resounding yes. My search for a pattern and consequently, a relationship started like this:

x	y	$x+y$	$x-y$
1	0	1	1
2	1	3	1
3	2	5	1
4	3	7	1
5	4	9	1

Table 11.1: x - y = 1

Let x = the nth option.

y = number of individual payments promised to be made by a customer.

11.9 Subtracting Number of Payments from Corresponding nth Option

Let us interpret **table 11.1** as follows:

1. Each time you subtract the number of payments from the corresponding n^{th} option, the result is always equal to 1.

Symbolically, $x - y = 1$.
Next is the graph of $y = x - 1$.

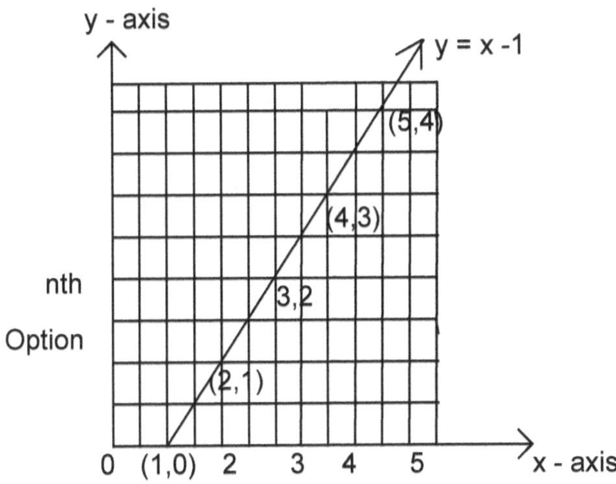

Figure 11.1: Graph of $y = x - 1$

11.10 Adding nth Option and Number of Payments

In each case, the nth option when added to the corresponding number of payments already made by a customer, the result is always an odd number. When we consider the options collectively, instead of individually, the result is the set of the numbers 1, 3, 5, 7, 9 which as we know forms a pattern.

$1 + 0 = 1 = 1 + (1\text{-}1) = \mathbf{1}$
$2 + 1 = 3 = 2 + (2\text{-}1) = \mathbf{3}$
$3 + 2 = 5 = 3 + (3\text{-}1) = \mathbf{5}$
$4 + 3 = 7 = 4 + (4\text{-}1) = \mathbf{7}$
$5 + 4 = 9 = 5 + (5\text{-}1) = \mathbf{9}$

Generally, $x + (x - 1) = 2x - 1 = x + (x - 1) = x + y$
Therefore, $x + y = 2x - 1 \Leftrightarrow x - y = 1$.
From here, if $x - y = 1$, then $y = x - 1$.
Next is the graph of $y = x - 1$.

x	y	x+y
1	0	1
2	1	3
3	2	5
4	3	7
5	4	9

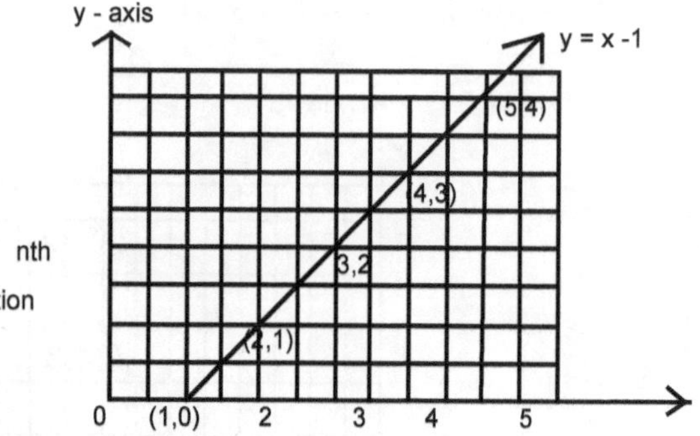

Figure 11.2: Graph of $y = x - 1$

11.11 The Result is Always Equal to One

2. When you add 2 to the number of payments already made by a customer, and subtract the number of options, the result is always equal to 1. Take a look!

y	x	$y + 2$	$(y + 2) - x$
0	1	2	1
1	2	3	1
2	3	4	1
3	4	5	1
4	5	6	1

Table 11.2: $(y + 2) - x = 1$

(2+0) - 1 = 1
(2+1) - 2 = 1
(2+2) - 3 = 1
(2+3) - 4 = 1
(2+4) - 5 = 1

Generally (2+y) - x = 1, y + 1 = x.

If y+1 = x, then from here, y = x - 1.

Next is the graph of y = x - 1.

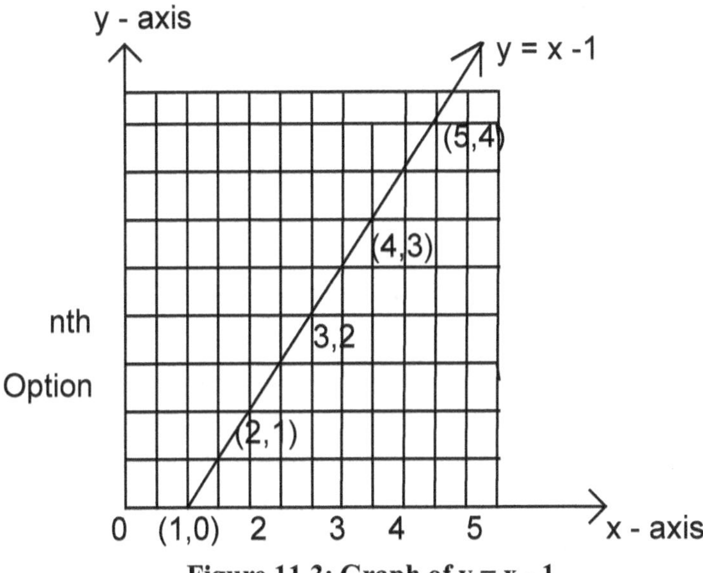

Figure 11.3: Graph of y = x - 1

Again, the same equation and the graph is the same.

11.12 Subtracting One from the nth Option

4. Twice the nth option minus 1 is equal to a set of odd numbers whose minimum value is 1 and the maximum value is 9.

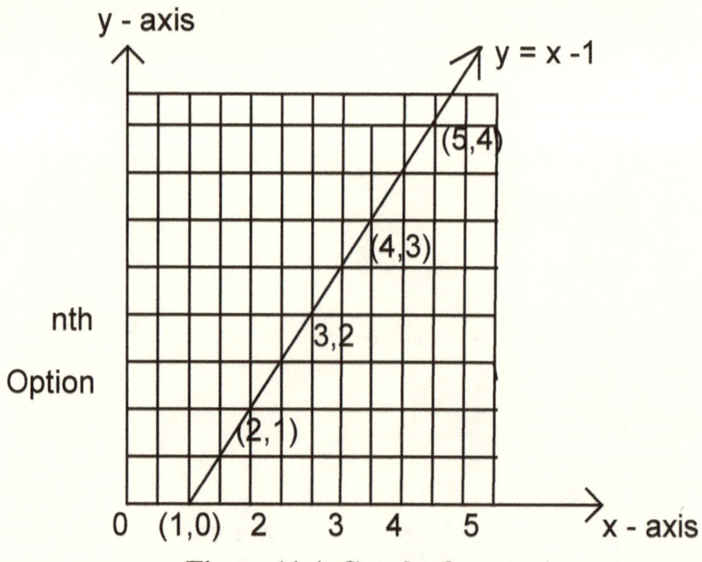

Figure 11.4: Graph of $y = x - 1$

Again, we have the same equation and the same graph. Amazing! Isn't it?

11.13 The Sum of nth and Fourth Options

5. The sum of twice the nth option, and the fourth option is equal to the sum of the nth option, corresponding number of equal payments made by a customer, and the total number of available options.

x	$2x + 4$	y	x+y	$x + y + 5$
1	6	0	1	6
2	8	1	3	8
3	10	2	5	10
4	12	3	7	12
5	14	4	9	14

Table 11.3: 2x+4 = x + y +5

Generally speaking, 2x + 4 = x + y + 5.
From here, 2x - x - y + 4 - 5 = 0.

This reduces to x-y-1 = 0 so that x-y = 1.
Next is the graph of y = x - 1.

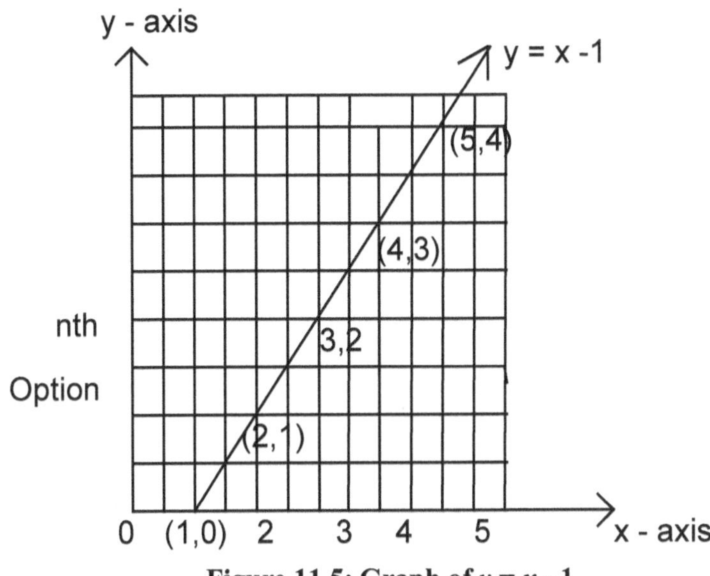

Figure 11.5: Graph of $y = x - 1$

WOW!! Again, the same equation and the same graph, $y = x - 1$.

Payment Options	*Number of Payments*
$\frac{1}{4}$	0
$\frac{2}{4}$	1
$\frac{3}{4}$	2
$\frac{4}{4}$	3
$\frac{5}{4}$	4

Table 11.4: Payment options and number of payments

11.14 Building a Set of Two Digit Numbers

Let us build a set of two digit numbers (call each xy) from table 11.5 below where x = the nth option and y = number of equal payments promised by a customer using the nth option.

Doing so, the possible values of xy in ascending order are 10, 21, 32, 43, and 54.

Look at the following subtraction facts:

x	y	$x_n y_n$	$x_{n+1} y_{n+1} - x_n y_n$
1	0	10	11
2	1	21	11
3	2	32	11
4	3	43	11
5	4	54	11

Table 11.5

The result of each subtraction fact is 11. We can represent these facts in a general sense and then strive to derive the relationship, y = x - 1.
From here, y = x - 1. The same equation.
Each of these subtraction facts can be represented generally as:

$$\left(\frac{y+1}{4}\right) - \frac{1}{4} = \frac{x-1}{4}.$$

Clearing fractions in the above equation we have:

$$\left(\frac{y+1}{4}\right)4 - \frac{1}{4}(4) = \left(\frac{x-1}{4}\right)4.$$

y + 1 - 1 = x - 1 $\Leftrightarrow$ y = x - 1.
From here, x - y = 1 $\Leftrightarrow$ y = x - 1
Next is the graph of y = x - 1.

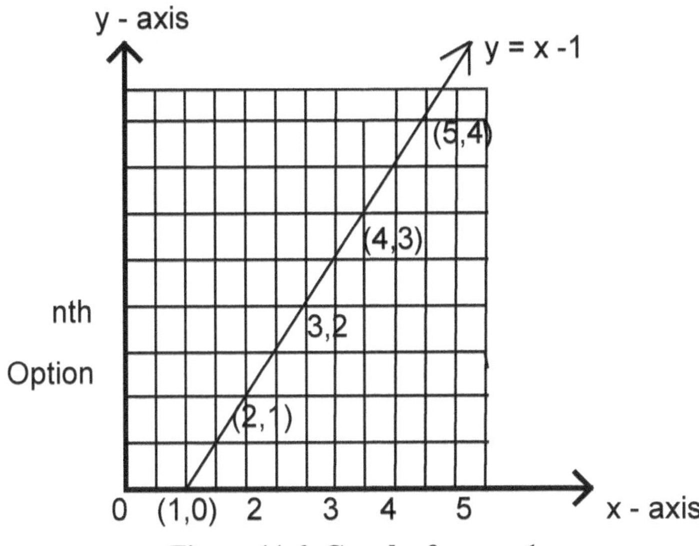

Figure 11.6: Graph of y = x - 1

Again, the same equation and the same graph.

11.15 Number of Payment Options as a Sum and a Difference

1. The sum of an nth option and 4 minus the number of payment options is equal to number of individual payments to be made by a customer. From here, we can derive the same equation, and consequently, the same graph.

x	$x + 4$	$(x + 4) - 5$	y
1	5	5 - 5 = 0	0
2	6	6 - 5 = 1	1
3	7	7 - 5 = 2	2
4	8	8 - 5 = 3	3
5	9	9 - 5 = 4	4

Table 11.6

(4+1) - 5 = 0	(4+2) - 5 = 1	(4+3) - 5 = 2
(4+4) - 5 = 3	(4+5) - 5 = 4	(4+6) - 5 =5

Table 11.7

Transforming the above into an equation we have:
$(4 + x) - 5 = y \Leftrightarrow x + (4 - 5) = y \Leftrightarrow x - 1$ or $y = x - 1$

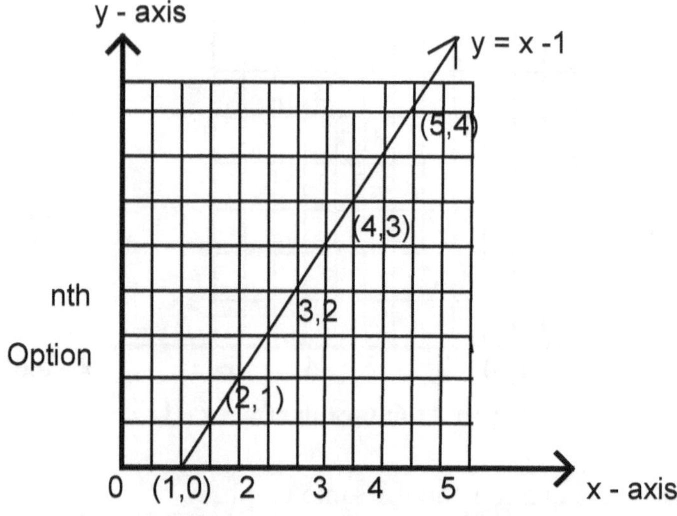

Figure 11.8: Graph of y = x - 1

11.16 Using Numerator and Denominator to Form a Two-Digit Number

8) Let us use each numerator and denominator to form a two-digit number whose tens digit are 1, 2, 3, 4, and 5 and ones digits as 4 in each case. Doing so, the two digit numbers are 14, 24, 34, 44, and 54.

x	y	$x_n y_n$	$x_{n+1}y_{n+1} - x_n y_n$
1	0	14	10
2	1	24	10
3	2	34	10
4	3	44	10
5	4	54	10

Table 11.8

24 - 14 = 10, 34 - 24 = 10, 44 - 34 = 10, 54 - 44 = 10
Each subtraction fact can be represented as:

$$[10(x+1)+4-(y+1)(10+4)+4=10]$$

Simplifying the above general form of subtraction fact we have:

10x + 10 + 4 - 10y - 10 - 4 = 10
10x - 10y +10 - 10 + 4 - 4 = 10
10x - 10y = 10 10(x-y)=10

Dividing throughout by 10 we have:

$$\frac{10(x-y)}{10} = \frac{10}{10} \Leftrightarrow x-y=1$$

From here, x - y = 1 $\Leftrightarrow$ y = x - 1.
The next is the graph of y = x - 1.

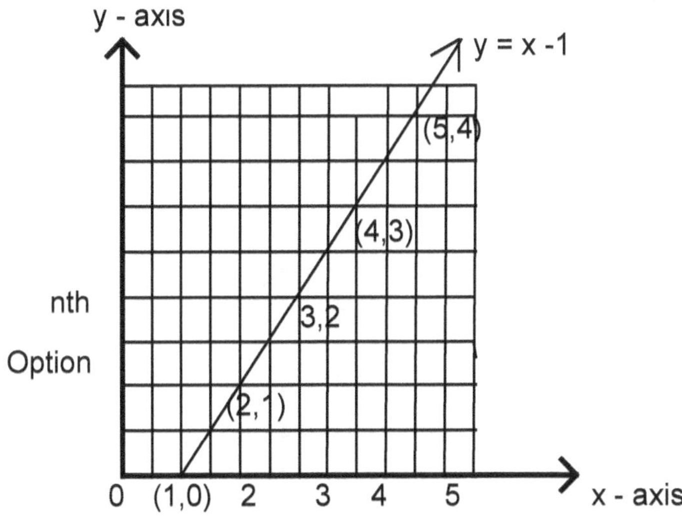

Figure 11.8: Graph of y = x - 1

Again, the following are true:

x	y	$x_n y_n$	$x_{n+1}y_{n+1} - x_n y_n$
1	0	01	11
2	1	12	11

3	2	23	11
4		34	11
5		45	11

Table 11.9

12 - 01 = 11, 23 - 12 = 11, 34 - 23 = 11, 45 - 34 = 11

11.17 Generating the Linear Equation, y = x - 1

9) Each of the above subtraction facts can be written generally as:

$$10x + (x+1) - (10y + x) = 11$$

$$= 10x - 10y = 10$$

Dividing both sides by 10 we have:

$\Leftrightarrow x - y = 1 \Leftrightarrow y = x - 1$

Next is the graph of y = x - 1.

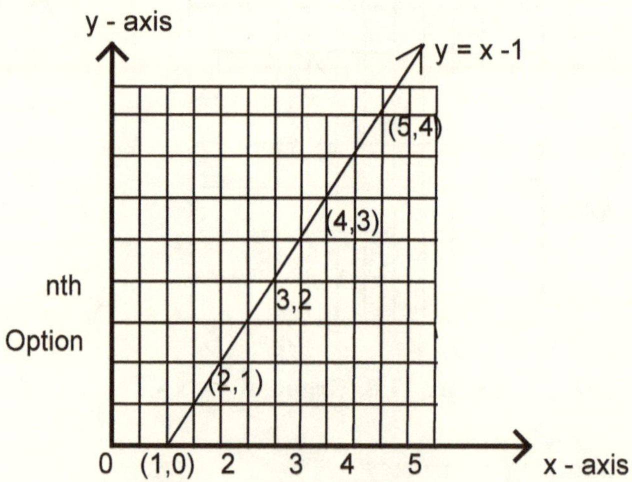

Figure 11.9: Graph of y = x - 1

Again, the same equation and the same graph.
Is this not amazing? Yes, it is!

11.18 Mathematical Significance

Where are we going with all this? What has algebra got to do with payment options and number of individual payments made by a customer? Do the patterns have any mathematical significance? The presence of these patterns suggests consistency. Consistency suggests fairness. Fairness suggests customer satisfaction. Customer satisfaction suggests progressive customer service and progressive customer service is the goal at Matrixx Marketing. What do you think?

Now take a look at related problems.

EXAMPLE 1: (i) The graph of the relation $y = mx + c$ where m and n are constants cuts the x-axis where x =1 and the y-axis where y = -1. Calculate the values of m and c. (ii) Using your values of m and c, solve the equation, $x^2 + mx + c = 0$

SOLUTION When $x = 1, y = 0$.

By substituting in $y = mx + c$ we have:
$0 = m, 1 + c \Leftrightarrow 0 = m + c$ ....(i)
Also, when $x = 0, y = 0$.
By substituting in $y = mx + c$ we have:
$-1 = 0, m + c \Leftrightarrow c = -1$
By substituting in equation (i) we have:
$0 = m + (-1)$.
From here, $0 = m - 1 \Leftrightarrow m = 1$.

Therefore, $m = 1$ and $c = -1$.
By substitution, we have:

$$x = \frac{-1 \pm \sqrt{1^2 - 4 \times 1 \times (-1)}}{2}$$

$$\frac{1 + \sqrt{5}}{2} = \frac{1 \pm 2.2361}{2} = 1.6180 \; or \; -0.6180$$

EXAMPLE 2 The graph of y = $x^2 + mx + c$ where m and c are constants cuts the x axis at the point (3,0) and the y-axis at the point (0,6). Find the values of m and c. **(S.C./G.C.E 1982)**

SOLUTION $y = mx + c$
When $x = 3$, $y = 0$ and when $x = 0$, $y = 6$
By substituting in $y = mx + c$,
$0 = 9 + 3m + c$
Also, $6 = 0 + c$.
From the above equation $c = 6$.
By substituting for c in $0 = 9 + 3m + c$, we have
$9 + 3m + 6 = 0 \Leftrightarrow m = -5$.
Therefore, $m = -5$.
Answer $m = -5$, $c = 6$.

EXAMPLE 3 (i) The graph of the relation $y = mx + c$ where m and n are constants cuts the x-axis where $x = 3$ and the y-axis where $y = 6$. Calculate the values of m and c. (ii) Using your values of m and c, solve the equation, $x^2\ mx + c = 0$. **(S.C./G.C.E 1980)**

SOLUTION When $x = 3$, $y = 0$.
By substitution in $y = mx + c$ for m and c we have:
$3^2 + 3m + c\ = 9 + 3m + c = 0$
From here, $3m + c = -9$
Also, when $y = 6$, $x = 0$.
By substituting in $y = mx + c$, we have:
$6 = 0 + 0 + c$.
From here, $x = 6$.
Substituting in $3m + c = -9$, we have:
$3m + 6 = -9$, $3m = -15$, $= \dfrac{-15}{3} = -5$
Answers: $m = 5$, $c = 6$

(ii) Substituting for the values of m and c in the relation, $x^2 + mx + c$ we have:
$x^2 + mx + c = x^2 + (-5x) + 6 = 0$

Factorizing we have:

$$x^2 - 5x + 6 = (x-2)(x-3) = 0$$

Therefore, x = 2 or 3.

11.19 A Search for Counter Examples

Can you identify any straight line graph that cannot be generally represented by the relation, $y = mx + c$?

Chapter 12

Making Connections within the Curriculum

12.1 Objectives

At the end of the lesson, the students should be able to:

(a) Link algebra with algebra
(b) Find the area of a triangle using Hero's formula
(c) Use the cosine formula to find the unknown sides a triangle
(d) Use the sine rule to find the unknown sides of a triangle
(e) Verify the area of a triangle using Hero's formula
(f) Find the area of a triangle using the formula: *Area* = ½ (*base* · *height*)
(g) State the Pythagorean theorem

12.2 Introduction

A posting I saw on the internet made me realize that finding the formula for the area of a triangle as half the area of a rectangle on the same base and equal height has more implications for the classroom than, perhaps, originally thought.

A high school math teacher has posted the following to a newsgroup on the internet asking how he/she could link geometry and algebra.

Date: 24 November 2000:

Hello everyone,

I was wondering if anyone could help me with a math problem I am having. I am wanting to know if there is any way that I can teach something in algebra and then verify or show how it works with geometry. Thank you for your time and contribution.

The above posting generated some interest in me and I decided to attempt a contribution. What follows is a result of that attempt.

Can we really link geometry with algebra? Yes, of course. In terms of the *National Curriculum and Evaluation Standards for High School Mathematics (*a document of the National Council of Teachers of Mathematics) that is a part of what connection is all about - the ability to link within the curriculum or outside the curriculum. In our case, it is linking within the curriculum - linking algebra with algebra.

12.3 Connecting Algebra with Algebra

The idea of deriving the general formula for the area of a triangle in terms of the base and the perpendicular height from a rectangle of equal base and equal height could be a good example of linking geometry and algebra. If this linkage is like drawing a circle, that circle is complete when we have been able to identify methods for finding the area of a triangle that depends solely on algebra for its investigation and logical understanding and then being able to bring the verification to a right conclusion. This verification takes the form of a general proof followed by examples.

One such method is Hero's formula. Hero's formula enables us to find the area of a triangle in terms of its sides a, b, and c. The linkage involves investigations that could help us to verify the area of a triangle as being equal to half base multiplied by the perpendicular height.

The linkage is therefore, not complete if we fail to identify at least one such example that could help us to bring the investigation to a logical conclusion. Therefore, if an algebraic verification does not take place, there is no connection. Having found the formula for the area of a triangle, we can now verify our result hoping that Hero's formula will help us do this. This verification, it is hoped, will complete the linkage.

12.4 Area of Triangle Using Hero's Formula

This formula enables us to find the area of a triangle in terms of its sides a, b, and c.

Our result is that in general, the area of a triangle is given by

$$\tfrac{1}{2}\,(\textit{base} \cdot \textit{perpendicular height})$$

To verify our result we need to go through the following steps:

(a) Find the following: s, s - a, s-b, s-c by substituting for s, a, b, c and simplify resulting expression..

(b) Substitute for s, s-a, s-b, s-c in

$$\sqrt{s(s-a)(s-b)(s-c)}.$$

(c) Evaluate the expression in (b) after substitution.

(d) If correctly worked out, the quantity, $\sqrt{s(s-a)(s-b)(s-c)}$ will evaluate to $\frac{1}{2}ab$ which is the expression for the area of a triangle in terms of its base and perpendicular height.

(e) Give examples of finding area of a triangle using Hero's formula.

Finding $s - a, s - b, s - c$ by substituting for s, given that $s = \dfrac{a+b+c}{2}$.

$$s - a = \frac{(a+b+c)}{2} - a \;=\; \frac{(a+b+c-2a)}{2} = \frac{c-a+b}{2}$$

$$s - b = \frac{(a+b+c)}{2} - b \;=\; \frac{(a+b+c-2b)}{2} = \frac{a+c-b}{2}$$

$$s - c = \frac{(a+b+c)}{2} - c \;=\; \frac{(a+b+c-2c)}{2} = \frac{a+b-c}{2}$$

(f) Substituting for s, s-c, s-b s-a in $\sqrt{s(s-a)(s-b)(s-c)}$ we have:

$$\text{Area of } \Delta\text{ABC} = \sqrt{\left(\frac{a+b+c}{2}\right)\left(\frac{a+b-c}{2}\right)\left(\frac{a+c-b}{2}\right)\left(\frac{c-a+b}{2}\right)}$$

(g) Evaluate the expression in (f). Doing so we have:

$$= \sqrt{\left(\frac{a+b+c}{2}\right)\left(\frac{a+b-c}{2}\right)\left(\frac{a+c-b}{2}\right)\left(\frac{c-a+b}{2}\right)}$$

$$= \sqrt{\frac{(a+b+c)(a+b-c)(a+c-b)(c-a+b)}{16}}$$

$$= \sqrt{\frac{(a^2+ab-ac)(ab+b^2-bc)(ac+bc-c^2)(c+a-b)(c-a+b)}{16}}$$

$$= \sqrt{\frac{(a^2+b^2)(ab+ab-ac+ac-bc-c^2)(c+a-b)(c-a+b)}{16}}$$

$$= \frac{\sqrt{2ab\left[(c+a-b)(c-a+b)\right]}}{16}$$

$$= \frac{\sqrt{2ab\left[(c^2-ac+bc+ac-a^2+ab-bc+ab-b^2)\right]}}{16}$$

$$= \frac{\sqrt{2ab\left[\left(c^2 - a^2\right) - b^2 - ac + ac + bc - bc + \left(ab + ab\right)\right]}}{16}$$

$$= \sqrt{\frac{(2ab)(2ab)}{16}}$$

Therefore, $\sqrt{s(s-a)(s-b)(s-c)}$

$$= \sqrt{\frac{4a^2b^2}{16}} = \sqrt{\frac{a^2b^2}{4}} = \frac{ab}{2} = \tfrac{1}{2}ab$$

Therefore, given that

$$s = \frac{a+b+c}{2},$$

$\sqrt{s(s-a)(s-b)(s-c)}$ is equal to $\tfrac{1}{2}ab$.

Give examples of finding area of a triangle using Hero's formula

When $\tfrac{1}{2}ab$ is translated into words it becomes half base multiplied by the perpendicular height which is the formula for the area of triangle.

We have therefore, used Hero's formula to verify that the area of a triangle is given by half base multiplied by the perpendicular height of the triangle and it does not matter whether the triangle is right-angled, acute-angled, or obtuse angled.

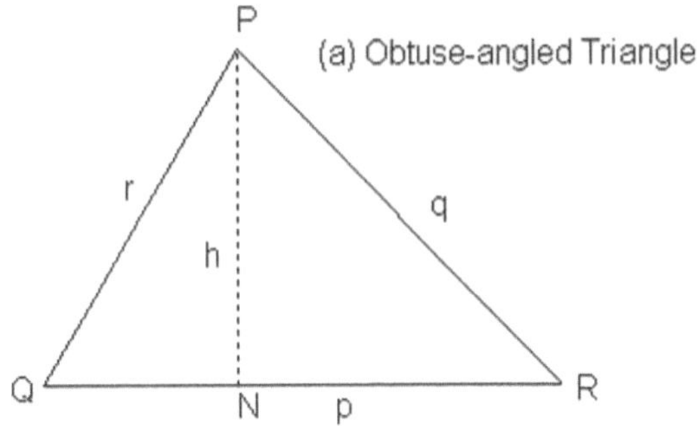

Area of $\Delta PQR = \sqrt{s(s-p)(s-q)(s-r)}$.

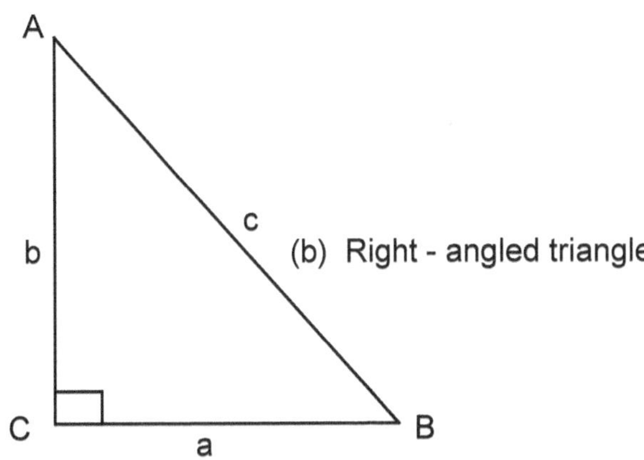

Area of $\Delta ABC = \sqrt{s(s-a)(s-b)(s-c)}$.

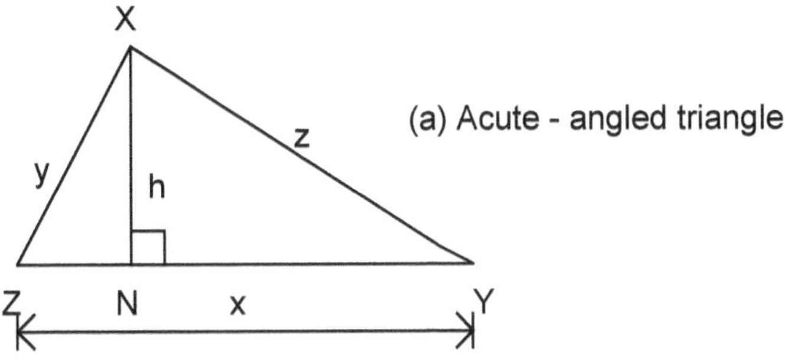

Area of $\Delta XYZ = \sqrt{s(s-x)(s-y)(s-z)}$.

Using the Cosine Rule to Find Unknown Sides

Consider ΔABC as shown below. If one of the sides is not given, we can use the cosine rule to find the unknown side.

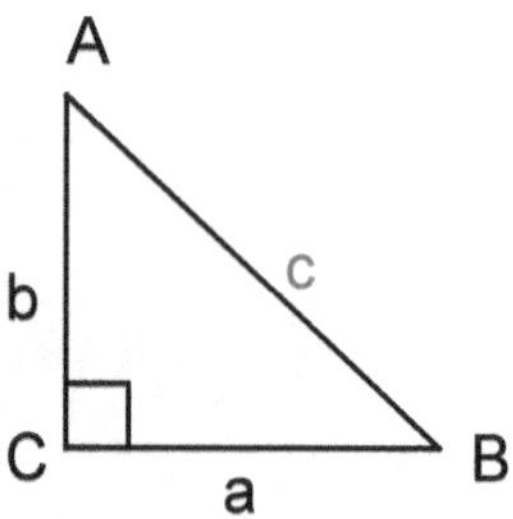

$$Cos\ \ A = \frac{b^2 + c^2 - a^2}{2bc}$$

$$Cos\ \ B = \frac{a^2 + c^2 - b^2}{2ac}$$

$$Cos\ \ C = \frac{a^2 + b^2 - c^2}{2ab}$$

If for example, $C = 90^\circ$, then

$$0 = \frac{a^2 + b^2 - c^2}{ab} \Leftrightarrow a^2 + b^2 = c^2, \quad \text{since Cos } 90^\circ = 0,$$

This demonstrates that the cosine rule is in fact, a modification of the Pythagorean theorem which states that in a right-angled triangle, the square on the hypotenuse is equal to the sum of the squares on the other two sides.

12.5 Using the Sine Rule to Find Unknown Sides

We can also use the sine rule for triangles to find the unknown elements.

$$\frac{a}{\sin A} = \frac{b}{\sin B} = \frac{c}{\sin C}$$

$$\Leftrightarrow a = \frac{b \sin A}{\sin B} = \frac{c \sin A}{\sin C}$$

$$\Leftrightarrow b = \frac{c \sin B}{\sin C} = \frac{a \sin B}{\sin A}$$

$$\Leftrightarrow c = \frac{b \sin C}{\sin B} = \frac{a \sin C}{\sin A}$$

With this we can find the unknown elements and proceed with Hero's formula since *a, b, c* are unknown at this point.

12.6 Finding Area of a Triangle Using Hero's Formula

***EXAMPLE 1*:** Without using Hero's formula, find the area of ΔABC with BC = AC,

SOLUTION

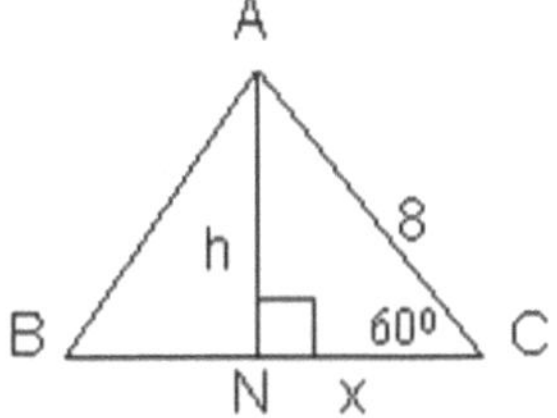

From A draw an altitude to meet BC at the point N.

Since AB = AC, N is the midpoint of BC.
Therefore, BN = CN.

Now let CN = x. With the notation in the figure, $h/8 = \sin 60^\circ$.

$$h = 8\sin 60^\circ = 8\left(\frac{\sqrt{3}}{2}\right) = 4\sqrt{3}$$

With the notation in the figure, in ΔANC, $x/8 = \sin 30^\circ$

$$x = 8\sin 30^\circ = 4$$

If x = 4, then BC = 8 (since BN = CN)

$$\text{Area of } \Delta ABC = \frac{1}{2}(\text{BC} \times \text{AN}) = \frac{1}{2}(\text{BC} \times h) = \frac{1}{2}(8 \times 4\sqrt{3})$$

$$= 16\sqrt{3} \text{ square units}$$

12.7 Verifying Our Result Using Hero's Formula

There are two ways we can do this:

(a) Using Hero's formula for the area of a triangle.

Finding the area of ΔACN. Since ΔABC is an equilateral triangle,

Area of ΔABC = 2 × area of ΔACN.

Since we are working with the same triangle, we have to find the unknown side by using the cosine rule.

Using Hero's formula we can verify the area.

$$a = 8,\ b = 8,\ s = \frac{a+b+c}{2}$$

12.8 Still Verifying the Area of a Triangle

Considering ΔACN, AC = 8, AN = $4\sqrt{3}$, and CN = 4

$$s = \frac{(8+4)+4\sqrt{3}}{2} = \frac{12+4\sqrt{3}}{2} = 6+2\sqrt{3}$$

$$s-a = (6+2\sqrt{3}) - 8 = 2\sqrt{3} - 2$$

$$s-b = (6+2\sqrt{3}) - 4\sqrt{3} = 6-2\sqrt{3}$$

$$s-c = (6+2\sqrt{3}) - 4 = 2 + 2\sqrt{3}$$

$$s(s-a)(s-b)(s-c)$$

$$= 6+2\sqrt{3}\left[(2\sqrt{3}+2)(2\sqrt{3}-2)(6-2\sqrt{3})\right]$$

$$= (6+2\sqrt{3})(6-2\sqrt{3})(2\sqrt{3}+2)(2\sqrt{3}-2)$$

$$= 6^2 - (2\sqrt{3})^2 \times \left[(2\sqrt{3})^2 - (2)^2\right]$$

$$= (36-12) \times (12-4)$$

$$= 24 \times 8 = 192$$

Area of $\Delta ACN = \sqrt{192} = \sqrt{16\times3\times4} = 8\sqrt{3}$
But area of $\Delta ABC = 2 \times \Delta CAN$.

Therefore, area of $\Delta ABC = 2\times8\sqrt{3} = 16\sqrt{3}$.

EXAMPLE 2

(a) Simplify $\left(1\frac{5}{8} + 1\frac{3}{5}\right) \div 5\frac{3}{8}$

(b) The sides of a triangle are 4 cm 4 cm and 4.8 cm. Calculate the area **(SC/GCE 1978).**

SOLUTION

(a) $\left(1\frac{5}{8} + 1\frac{3}{8}\right) \div 5\frac{3}{8} = \left(\frac{13}{8} + \frac{8}{5}\right) \times \frac{8}{43}$

$$= \frac{65+64}{40} \times \frac{8}{43} = \frac{129}{40} = \frac{43\times3\times8}{43\times5\times8} = \frac{3}{8}$$

(b) Let the triangle be named XYZ and also let A = Area of ΔXYZ. XY = 4 cm, XZ = 4 cm, and YZ = 4.8 cm. Using Hero's formula,

By substitution, $A = \sqrt{6.4(6.4-4)(6.4-4)(6.4-4.8)}\ \text{cm}^2$

$= \sqrt{(6.4)(2.4)(2.4)(1.6)}\ \text{cm}^2$

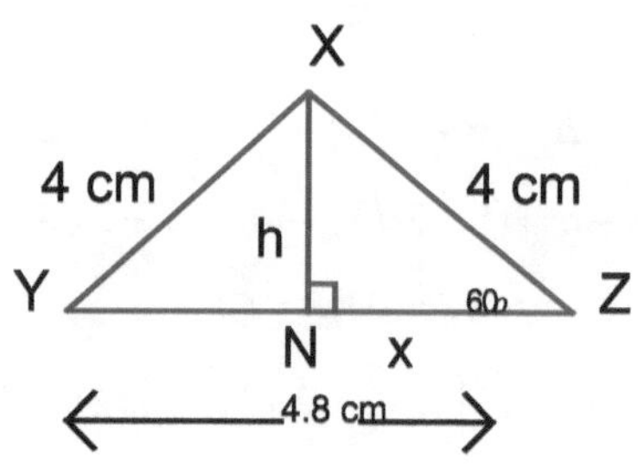

$= \sqrt{\dfrac{(64)(24)(24)(16)}{(10)(10)(10)(10)}}\ \text{cm}^2$

$= \dfrac{(8)(24)(4)}{(10)(10)} = \dfrac{768}{100} = 7.68\ \text{cm}^2$

Answer: Area of $\Delta XYZ = 7.68\ \text{cm}^2$.

12.9 Practice Exercises

(a) Using Hero's formula, prove that in an equilateral triangle, with sides s, the area is equal to $\frac{1}{4}\sqrt{3s^2}$ (**b**) In a right-angled triangle ABC, show that by using Hero's formula, the area of ΔABC is the same as using

Area = ½ (*base* · *height*)

12.10 A Search for Counter Examples

(a) Can you identify any triangle with a given base and height whose area (A) cannot be found by using the formula,

$A = \frac{1}{2}(Base \times Perpendicular\ Height)$?

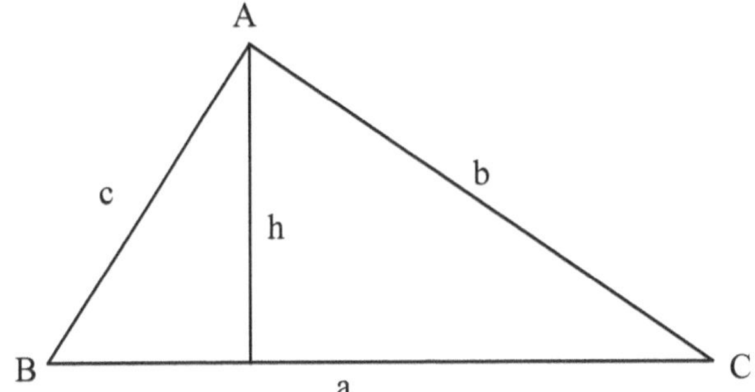

Can you identify any triangle ABC with sides BC = a, AC = b, and AB = c, whose area (A) cannot be found by using Hero's formula?

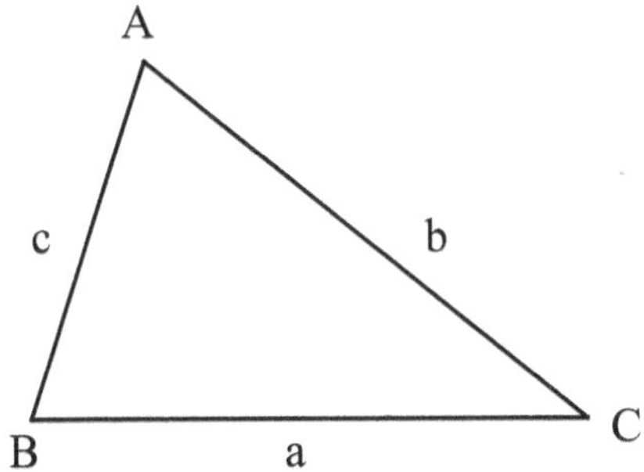

$A = \sqrt{s(s-a)(s-b)(s-c)}$, where $s = \dfrac{a+b+c}{2}$?

Chapter 13

Beyond Shapes: Triangular Numbers are More Than Just Shapes

13.1 Objectives

At the end of the lesson, the students should be able to:

(a) Give examples of triangular numbers
(b) Answer questions based on triangular numbers

13.2 Introduction

One evening as I was meditating on the contents of the National Council of Teachers of Mathematics Curriculum and Evaluation Standards for High School Mathematics, a thought flashed through my mind, a thought that relates to an article on triangular numbers I was writing at the time. It goes like this:

Imagine a school environment or a workplace, a group of seven students or employees as the case may be. One of the students or employees named Miki decided to be generous by providing free drinks for all the students or employees through the principal's or supervisor's office. She decided not to discuss this gesture with any of her fellow students or employees.

She also expressed to the principal or supervisor her desire to remain anonymous. In other words, she does not wish her name be disclosed to other students or employees as the one who provided the free drinks. Now each of the students or employees has decided to ask each other to find out who donated the drinks. They want to know. They couldn't hold it anymore.

Assume Miki at this point has decided to break her anonymity; each of the students will attract a *no* response from every other student, except Miki from whom every student or employee will attract a *yes* response because she is the one playing the Good Samaritan. Have you got it? Now go from here.

13.3 Unexpected Pattern

Initially, I was not serious about the investigation because I was not sure what to expect. But what I found about the total number of *no* responses was shocking. The total number of *no* responses presents a pattern. At this time, I concentrated only on the total number of *no* responses. Surprisingly, the total number of *yes* responses also presented a pattern. Here is the rest of what I came up with.

13.4 Type and Direction of Response

Assuming *A, B, C,* and *D* represent individual students or employees otherwise known as respondents. The response pattern for individual situations is shown on a table and each table has three or four columns.

1) List of Respondent column
2) Direction of Response column
3) Type of Response column
4) Number of *no* Response column
5) Number of *yes* Response column

Type of Response here means whether a particular response is either *yes* or *no*.

The arrow originates from the individual making the *no* or *yes* response.

A → B could mean a *no* response from A to B or yes response from A to B.

A → B under the *no* response column means a *no* response from A to B while

B → A in the same column means a no response from B to A.

Similarly, B → A in the *yes* response column means a *yes* response from B to A.

13.5 Pattern of Response with Two Respondents

Respondents	Type of Response	Direction of Response	Number of Responses
A	No	A → B	1
B	Yes	B → A	1

Figure 13.1: Direction of Response

List of Respondents	*no* Responses	*yes* Responses
A	1	0
B	0	1
Total	1	1

Table 13.1: Summary of Responses by A, B

13.6 Pattern of Response with Three Respondents

Respondents	Type of Response	Direction of Response	Number of Responses
A	No	A → B A → C	1 1

B	No	B→A B→C	1 1
C	Yes	C→A C→B	1 1

Figure 13.2: Direction of Response by A, B, C

The above can be interpreted as follows:

A will say no to B
A will say no to C
B will say no to A
B will say no to C

Total number of *no* responses = 4 = 2^2.

C will say yes to A and C will say yes to B.

Total number of *yes* responses = 2.

Summary of Responses by A, B, C only

List of Respondents	*no* Responses	*yes* Responses
A B C	2 2 0	0 0 2
Total	4	2

Table 13.2: Summary of Responses by A, B, C

13.7 Pattern of Response with Four Respondents

Respondents (ABCD)	Type of Response	Direction of Response	No. of *No* Responses
A	No No No	A→B A→C A→D	**1** **1** **1**
B	No No No	B→A B→C B→D	**1** **1** **1**
C	No No No	C→A C→B C→D	**1** **1** **1**
Respondents	Yes Yes Yes	**Direction of Response**	**No. of yes Responses**
D	Yes Yes Yes	D→A D→B D→C	1 1 1

Figure 13.3: Direction of Response by A, B, C, D

A will say no to B B will say no to A C will say no to A D will say yes to A
A will say no to C B will say no to C C will say no to B D will say yes to B
A will say no to D B will say no go C C will say no to D D will say yes to C

Total number of *no* responses = 9 = 3^2

Total number of *yes* responses = 3.

List of Respondents	No. of no Responses	No. of yes Responses
A	3	0
B	3	0
C	3	0
D	0	3
Total	9	3

Table 13.3: Summary of Responses by A, B, C, D

13.8 Pattern of Response with Five Respondents

Respondents	Type of Response	Direction of Response	Responses
A	No	$A \rightarrow B$	1
	No	$A \rightarrow C$	1
	No	$A \rightarrow D$	1
	No	$A \rightarrow E$	1
B	No	$B \rightarrow A$	1
	No	$B \rightarrow C$	1
	No	$B \rightarrow D$	1
	No	$B \rightarrow E$	1
C	No	$C \rightarrow A$	1
	No	$C \rightarrow B$	1
	No	$C \rightarrow D$	1
	No	$C \rightarrow E$	1
D	No	$D \rightarrow A$	1
	No	$D \rightarrow B$	1
	No	$D \rightarrow C$	1
	No	$D \rightarrow E$	1
E	Yes	$E \rightarrow A$	1
	Yes	$E \rightarrow B$	1
	Yes	$E \rightarrow C$	1
	Yes	$E \rightarrow D$	1

Figure 13.4: Direction of Response by A, B, C, D, E

The above table can be interpreted as follows:

A will say no to B A will say no to C A will say no to D A will say no to E		A will say no to B A will say no to C A will say no to D A will say no to E		A will say no to B A will say no to C A will say no to D A will say no to E

A will say no to B A will say no to C A will say no to D A will say no to E		A will say no to B A will say no to C A will say no to D A will say no to E

Responses by A, B, C, D, and e only

Total number of no responses = 16 = 4^2

Table 14 is for *no* responses. Now for the *yes* responses.

E will say yes to A E will say yes to B E will say yes to C E will say yes to D

Table 13.4: Responses by E only

Total number of *"yes"* responses = 4.

List of Respondents	*no* Responses	*yes* Responses
A	4	0
B	4	0
C	4	0
D	4	0
E	0	4
Total	16	4

Table 13.5: Summary of Responses by A, B, C, D, E

List of Respondents	*no* Responses	*yes* Responses
AB	1	1
ABC	4	2
ABCD	9	3
ABCDE	16	4

Table 13.6: Summary of Responses by A, B, C, D, E

13.9 Still in Search of the nth Triangular Number

From our investigation so far, it seems that there are infinite number of respondents. Actually, there are. In order to elicit at least a *no* response and a *yes* response, there must be at least two respondents. The number of respondents therefore, comprises of the set of numbers:

2, 3, 4, 5, 6, 7, 8, 9, 10, 11.

Respondents	*no* Responses	*yes* Responses
2	1	1
3	4	2
4	9	3
5	16	4
7	36	6

Table 13.7: Finding the nth Triangular Number

We can patiently complete this search as follows:

Let the number of *no* responses = x.

For each number of respondents, find:

(a) The number of *no* respondents in terms of *x*.
(b) The number of *yes* responses in terms of *x*.
(c) The sum of your result in (a) and (b) all in terms of x only. Call this p.
(d) Divide your result in (a) by 2. What do you notice?

You can defer step (d) until after step (f).

(e) Divide your result in (b) by 2. What do you notice?

You can also defer step (e) until after step (g).

(f) Express your result in (d) in terms of a only.
(g) Express your result in (e) in terms of n only.
(h) Find the sum of your result in (f) and (g).

What do you notice?

From figure 13.6, $x - n = 1$.

Therefore, $x = n + 1$ ....(i)

Therefore, $p = x^2 - x$ ....(ii)

Substituting for x in (ii) above we have:

$P = (n+1)^2 - (n+1)$

$= n^2 + 2n + 1 - (n+1) = n^2 + n$ ....(iii)

Therefore, $p = n^2 + n$.

x	y	z	$y+z$	$x+y+z$	xz	$x+y+z+xz$
2	1	1	2	4	2	6
3	4	2	6	9	6	15
4	9	3	12	16	12	28
5	16	4	20	25	20	45
6	25	5	30	36	30	66
7	36	6	42	49	42	91

Figure 13.5

n	*x*	*x-n*		*n*	*x*	*x-n*
1	2	1		7	8	1
2	3	1		8	9	1
3	4	1		9	10	1
4	5	1		10	11	1
5	6	1		11	12	1
6	7	1		12	13	1

Figure 13.6

13.10 Relationships between Responses

From figure 13.5, we can make the following observations:

The product of the number of *no* responses (call it y) and the number of *yes* responses (call it z) is equal to the cube of the number of *yes* responses.

$yz = x^3$

Symbolically, $yz = z^3$ and this demonstrates that the number of *no* responses is the square of the number of *yes* responses since $y = z^2$

$$yz = (x-1)^3 = (x^2 - 2x + 1)(x - 1)$$

$$(x^2 - 2x + 1)(x - 1) = x(x^2 - 2x + 1) - 1(x^2 - 2x + 1)$$

$$= x^2 - 2x^2 + x - \{x^2 - 2x + 1\}$$

$$= x^3 - 2x^2 + x - x^2 + 2x - 1$$

$$x^3 - 2x^2 - x^2 + x + 2x - 1$$

$$= x^3 - 3x^2 + 3x - 1 \text{ .... (Equation 1)}$$

But x = n+1. By substitution in equation 1 for x we have:

$$x^3 - 3x^2 + 3x - 1 = x^3 - 3x^2 + 3x - 1$$

$$= (n+1)^3 - 3(n+1)^2 + 3(n+1) - 1$$

$$= (n^2 + 2n + 1)(n+1) - 3(n+1)^2 + 3(n+1) - 1$$

$$n^3 + 2n^2 + n + n^2 + 2n + 1 - 3n^2 - 6n - 3 + 3n + 3 - 1$$

$$= n^3 + (2n^2 + n^2 - 3n^2) + (n + 2n + 3n) - 6n - 3 + 3n + 3 + 1 - 1$$

$$= n^3 + (3n^2 - 3n^2) + (6n - 6n) + (3 - 3) + (1 - 1) = n^3$$

Therefore, $yz = (x-1)^3 = n^3$.

If $(x-1)^3 = n^3$, taking the cube root of both sides we have:

$x - 1 = n \Leftrightarrow x = n + 1$.

13.11 Triangular Numbers as Half Sum

For any yes number of respondents, half the sum of the number of *no* responses and the number of *yes* responses is equal to a triangular number.

Above can be verified as follows:

If the number of respondents is represented by x, then the following are true:

Number of *no* responses = x - 1.

Number of *yes* responses = x - 1.

$$Now \quad \frac{y+z}{y} = \frac{(x-1) + x - 1}{2}$$

By substituting in equation we have:

$$\frac{y+z}{2}=\frac{(x-1)^2+x-1}{2}$$

$$\frac{(x-1)^2+x-1}{2}=\frac{x^2-2x+1+x-1}{2}$$

$$=\frac{x^2-x}{2}=\frac{x(x-1))}{2}$$

By substitution for x in $\frac{x(x-1))}{2}$ we have:

$$\frac{x(x-1))}{2}=\frac{n+1[(n+1)-1]}{2}=\frac{n(n+1)}{2}$$

13.12 Triangular Numbers as Average of Two Products

Half the product of the number of respondents and the number of no responses and the number of yes responses is equal to a set of triangular numbers,

$x + y + z = x^2$

Again the sum of the number of respondents, number of no responses and yes responses is a set of the squares of the numbers 2, 3, 4, 5, 6, 7 . . . In other words, the sum of the number of respondents, number of *no* responses and the number of *yes* responses is always a square of the number of respondents.

$x + y + z = x^2$, $x = n + 1$, $y = (x - 1)^2$ and

$(x - 1)^2 = (x - 1)^2 = [(n + 1) - 1]^2$

$z = x - 1$ and $x - 1 = (n + 1) - 1 = n$

$x + y + z = n + 1 + n^2 + n$

$= n^2 + 2n + 1 = (n + 1)^2$.

But $n + 1 = x$ ....(ii)

Therefore, $(n+1)^2 = x^2$.

Consequently, $x + y + z = x^2$ and x^2 is the number of respondents.

13.13 Responses Expressed in Factorial Notation

Search of a general expression for *no* and *yes* number of responses

(a)

$1 \text{ x} \frac{(2-1)!}{(2-2)!} = 1 \text{ x } \frac{1!}{0!} = 1$ (since $0! = 1$)	$3 \text{ x} \frac{(4-1)!}{(4-2)!} = 9$
$2 \text{ x} \frac{(3-1)!}{(3-2)!} = 4$	$4 \text{ x} \frac{(5-1)!}{(5-2)!} = 16$

(b)

$5 \text{ x} \frac{(6-1)!}{(6-2)!} = 25$	$7 \text{ x} \frac{(8-1)!}{(8-2)!} = 49$
$6 \text{ x} \frac{(7-1)!}{(7-2)!} = 36$	$8 \text{ x} \frac{(9-1)!}{(9-2)!} = 64$

(c)

$9 \text{ x} \frac{(10-1)!}{(10-2)!} = 81$	$11 \text{ x} \frac{(12-1)!}{(12-2)!} = 121$
$10 \text{ x} \frac{(11-1)!}{(11-2)!} = 100$	$12 \text{ x} \frac{(13-1)!}{(13-2)!} = 144$

There is surely a pattern here. Have you seen it?

Each of the above expressions represents the number of *no* responses.

Now study the above patterns and find a general expression for the number of *no* responses in factorial notation.

a) Leave your answer in the form $\frac{a}{b}$ where a and b are in factorial forms.
b) Leave your answer in its simplest form.
c) Why is your answer in (b) special?

Now study the above patterns and find a general expression for the number of *no* responses in factorial notation.

13.14 A Search for Counter Examples

(a) Can you identify any triangular number, T_n such that

$$T_n \neq \frac{n(n+1)}{2}?$$

(b) Can you identify any triangular number that cannot be expressed in factorial notation?

Chapter 14

A Student's Logic Under Trial: Verifying a Summation Strategy for the First n Fibonacci Numbers

14.1 Objectives

At the end of the lesson, the students should be able to:

(a) realize why it is wrong to state $(u_{2n+1} - 1) + (u_{2n})$ as sum (S_n) of first n Fibonacci numbers

(b) validate $u_{n+2} - 1$ as sum of first n Fibonacci numbers

14.2 Introduction

In this chapter a student, Master Emeka Ibezimako proposes a summation strategy for first n Fibonacci numbers. Here, we will state what the problem is. We will also provide and take a critical look at the student's solution. Next, we will also verify the correctness of $u_{n+2} - 1$ as the sum of first n Fibonacci numbers. Finally, we will end the discussion with a conclusion that tells us whether Emeka is right or wrong.

14.3 Getting Started

Let us get started by considering sets A and B below.

A = {1, 3, 8, 21, 55, 144, 377, 987, 2584, 6765, 17711}
B = {1, 2, 5, 13, 34, 89, 233, 610, 1597, 4181, 10946, 28657}

The union of the above two sets yields a third set.

Call this set C. Stated differently, set C is the set of Fibonacci numbers whose first term is equal to 1 and the last term is 28657.

In other words, C = A∪B.

C = {1, 1, 2, 3, 5, 8, 13, 21, 34, 55, 89, 144, 233, 377, 610, 987, 1597, 2584, 4181, 6765, 17711, 28657}

14.4 The Student's Approach

Now a student wants to find the sum of the elements in set C. Since set C is the union of sets A and B, the summation of the elements in set C can be found by combining individual summation strategies for each set as listed below in (a) and (b), he had told himself.

(a) The sum of first n even-subscripted Fibonacci numbers is given by u_{2n+1} - 1.
(b) The sum of first n odd-subscripted Fibonacci numbers is given by u_{2n}.

then $S_n = \{13 + 8 + 21 + 55 + 144 + 377 + 987 + 2584 + 6765 + 17711\}$

$+ \{1 + 2 + 5 + 13 + 34 + 89 + 233 + 610 + 1597 + 4181 + 10946 + 28657\}$

Now regardless whether n is odd or even, for first n Fibonacci numbers:

$$S_n = u_{n+2} - 1$$

Also, for first n Fibonacci numbers, let us re-visit the following facts:

$S_n = u_{2n+1} - 1$ for first n even-subscripted Fibonacci numbers.

$S_n = u_{2n}$ for first n even-subscripted Fibonacci numbers.

14.5 What is Wrong With the Student's Reasoning?

Is there anything wrong with the expression,

$$(u_{2n+1} - 1) + u_{2n}$$

which is the student's strategy for finding the sum of first *n* Fibonacci numbers whose last term is 28657? Yes, there is something wrong. The student's goal is to correctly sum up and then verify his solution. He failed to do both.

14.6 Mixing Apples with Oranges

What Master Ibezimako did is analogous to mixing apples with oranges.

Look at the following statements:

3 apples + 5 apples = 6 apples .... (a)
3 oranges + 5 oranges = 5 oranges.. (b)

On the other hand, compare (a) and (b) with (c) and (d).

Does that make sense?

3 apples + 5 oranges = 8 oranges..(c)
3 apples + 5 oranges = 8 apples..(d)

If we can transform or change an apple into an orange or an orange into an apple in all aspects, then statements (c) and (d) above would have been false. Until then (when an orange can be changed into an apple and vice versa), (c) and (d) will remain true statements. There is therefore, no relationship between an apple and an orange. Ibezimako needs help. Help is coming. I promise.

14.7 Need for a Relationship between Different Sets

Unlike the case of apples and oranges, we can establish a relationship between a given number of first *n* Fibonacci numbers and the following:

(a) No. of even-subscripted Fibonacci numbers
(b) No. of odd-subscripted Fibonacci numbers

How we can do this comes next. First of all, there is a need for a new notation for (a) and (b) above. Intuitively, some readers may recognize this need, while some like Ibezimako may not.

14.8 Our Sincere Concern

Master Ibezimako put all this into consideration. Therefore, logically, he reasoned that he could find the sum of the terms in set C by just adding u_{2n} and $u_{2n+1} - 1$. That was exactly what he did and our concern is whether his action was logical or otherwise.

But $u_{2n+1} - 1 + u_{2n} \neq u_{2n+2} - 1$ or are they?

Is the student's logic right or wrong?

What do you think?

(a) When n is odd and $n(A) = n(B)$.

This can happen in two ways:

1) Either $n(A)$ is odd and $n(B)$ is even since odd + even = odd

EXAMPLE 1

If $A = \{ 1, 3, 8, 21, 55, 144, 377 \}$ and $B = \{ 1, 2, 5, 13, 34, 89 \}$, then $n(A) - n(B) = 1$

If n is odd, then the following are true about n:

(1) Number of consecutive even-subscripted Fibonacci numbers is $\frac{n+1}{2}$

(2) Number of consecutive odd-subscripted Fibonacci numbers whose first term is equal to 1 is $\frac{n+1}{2}$, so that $\frac{n-1}{2} + \frac{n+1}{2} = n$

or $n(A)$ is even and $n(B)$ is odd since even + odd = odd.

EXAMPLE: If $A = \{ 1, 3, 8, 21 \}$ and $B = \{ 1, 2, 5, 13, 34 \}$,

Then $n(B) - n(A) = 1$. If n is odd, and $n(B) - n(A) = 1$, then the following are true about n.

(a) Number of consecutive even-subscripted Fibonacci numbers whose first term is equal to 1 = $\frac{n-1}{2}$.

(ii) Number of consecutive even-subscripted Fibonacci numbers whose first term is equal to 1 = $\frac{n-1}{2}$ so that $\frac{n-1}{2} + \frac{n+1}{2} = n$.

Let this notation be equal to *k* for both (a) and (b) above.

When *n* is Odd

A) Finding the Number of Even-Subscripted Fibonacci Terms

First *n* Fibonacci Numbers	*n*	*k*	*n* - *k*	*k* + 1
1	1	0	1	1
1,1,2	3	1	2	2
1,1,2,3,5	5	2	3	3
1,1,2,3,5,8,13	7	3	4	4

Table 14.1: Finding Number of Even-Subscripted Fibonacci Numbers

Let k = No. of even-subscripted Fibonacci numbers, given an odd number of first n Fibonacci numbers.

From Table 14.1, n-k = k+1 $\Leftrightarrow 2k = n - 1 \Leftrightarrow k = \dfrac{n-1}{2}$.

Conclusion: For any odd number of first n Fibonacci numbers, there are $\dfrac{n-1}{2}$ even-subscripted terms.

When n is Odd

Finding the Number of Odd-Subscripted Fibonacci Terms

First n Fibonacci numbers	*n*	*k*	*n* - *k*	k - 1
1	1	1	0	0
1,1,2	3	2	1	1
1,1,2,3,5	5	3	2	2
1,1,2,3,5,8,13	7	4	3	3

Table 14.2: Finding Number of Odd-Subscripted Fibonacci Numbers

From table 14.2, n-k = k-1. From here, n+1 = 2k.

Therefore, k $= \frac{n+1}{2}$.

Conclusion: For any odd number of first n Fibonacci numbers, there are $\frac{n+1}{2}$ odd subscripted terms.

When n is Odd

Finding the Number of Odd-Subscripted Fibonacci Terms

First *n* Fibonacci numbers	*n*	*k*	2*k* - *n*
1	1	1	1
1,1,2	3	2	1
1,1,2,3,5	5	3	1
1,1,2,3,5,8,13	7	4	1

Table 14.3: Finding Number of Odd-Subscripted Fibonacci Numbers

From table 14.3, $2k - n = 1$ so $n = 2k - 1$.

From here, $k = \frac{n+1}{2}$.

Conclusion: For any odd number of first n Fibonacci numbers, there are $\frac{n+1}{2}$ odd-subscripted terms.

When n is Even

Finding the Number of Odd and Even-Subscripted Fibonacci Terms

First *n* Fibonacci numbers	*n*	*k*	*n* - 2*k*
1,1	2	1	0
1,1,2,3	4	2	0
1,1,2,3,5,8	6	3	0
1,1,2,3,5,8,13,21	8	4	0

Table 14.4: Finding Number of Even-Subscripted Fibonacci Numbers

From Table 14.4, n-2k = 0.

From here, n-2k = 0 $\Leftrightarrow$ $k = \frac{n}{2}$

Conclusion; For any given even number of first n Fibonacci numbers, there are $\frac{n}{2}$ even-subscripted terms and $\frac{n}{2}$ odd-subscripted terms.

When n is Even

Finding the Number of Odd and Even-Subscripted Fibonacci Terms

First *n* Fibonacci numbers	*n*	*k*	*n* - 2*k*
1,1	2	1	0
1,1,2,3	4	2	0
1,1,2,3,5,8	6	3	0
1,1,2,3,5,8,13,21	8	4	0

Table 14.5: Finding Number of Even-Subscripted Fibonacci Numbers

From Table 14.5, $n - k = k$.

From here, n = 2k $\Leftrightarrow$ $k = \frac{n}{2}$

Conclusion: For any given even number of first n Fibonacci numbers, there are $\frac{n}{2}$ even-subscripted terms and $\frac{n}{2}$ odd-subscripted terms.

When n is Even

Finding the Number of Odd and Even-Subscripted Fibonacci Terms

First *n* Fibonacci numbers	*n*	*k*	2*k* - *n*
1,1	2	1	0
1,1,1,2,3	4	2	0
1,1,2,3,5,8	6	3	0
1,1,2,3,5,8,13,21	8	4	0

Table 14.6: Finding Number of Odd and Even-Subscripted Fibonacci Numbers

From Table 14.6, $2k - n = 0$.

$$2k = n \Leftrightarrow k = \frac{n}{2}$$

Conclusion: For any given even number of Fibonacci numbers, there are $\frac{n}{2}$ odd-subscripted terms and $\frac{n}{2}$ even-subscripted terms since for odd and even cases of n, $k = \frac{n}{2}$.

From here, $2k = n$

n	k	***n* - *k***
2	1	1
4	2	2
6	3	3
8	4	4

From here, $2k - n = 0 \Leftrightarrow k = \frac{n}{2}$.

Conclusion: For any given even number of Fibonacci numbers, there are $\frac{n}{2}$ odd-subscripted terms and $\frac{n}{2}$ even-subscripted terms since for odd and even cases of n, $k = \frac{n}{2}$

14.10 Verifying the Correctness of $u_{n+2} - 1$

To verify the correctness of $S_n = u_{n+2} - 1$ as the sum of the first n Fibonacci numbers by adding $u_{2n+1} - 1$ and u_{2n}, the student is saying that $u_{2n+1} - 1 + u_{2n} = u_{n+2} - 1$.

Knowing what we know now, we know that this is not true. Actually, $\left(u_{2n+1} - 1\right) + u_{2n} = u_{2n+2} + 1$ and not $u_{n+2} - 1$.

What does this immediately suggest?

When n is Even

We have to remember that when n is even, we choose the number of elements in set A, call it x_1 to be equal to the number of elements in set B (call it x_2) so that

$$x_1 = \frac{n}{2} \text{ and } x_2 = \frac{n}{2}.$$

Then the number of elements in each set should be $\frac{n}{2}$ and not *n* as the student erroneously thought. If we are finding the sum of first *n* Fibonacci numbers regardless whether *n* is odd or even, then it becomes a sound logic to refer to the sum as u_{n+2} - 1.

When *n* is even, n(A) = n(B).

When this is the case, the following are true:

(a) No. of consecutive even-subscripted Fibonacci numbers whose first term is 1 is equal to $\frac{n}{2}$

(b) No. of consecutive even-subscripted Fibonacci numbers whose first term is 1 is equal to $\frac{n}{2}$

From (a) and (b) above,

$$\frac{n}{2} + \frac{n}{2} = n$$

$$\frac{n}{2} - \frac{n}{2} = 0$$

Then according to Mazi Ibezimako,

$$S_n = (u_{2n+1} - 1) + u_{2n}$$

To verify the correctness of $S_n = \left(u_{2n+1} - 1\right) + u_{2n}$ as sum of first n Fibonacci numbers, by adding $\left(u_{2n+1} - 1\right)$ and u_{2n}, Ibezimako is saying again that

$S_n = (u_{2n+2} - 1) + u_{2n}$.

Actually, $S_n = (u_{2n+2} - 1) + u_{2n} \neq u_{n+2} - 1$.

In addition, the implication of the student's reasoning is that

$S_n = (u_{2n+2} - 1) + u_{2n} = u_{n+2} - 1$ is a true statement.

Do I agree with Ibezimako? No, I don't.

What do you think?

What I think is not important.

What is important is what you the readers think.

When n is Odd

On the other hand, when n is odd,

$n(A) - n(B)$ or $n(B) - n(A) = 1$
or $n(B) - n(A) = 1$.

When that is the case, the following are true:

Number of consecutive odd-subscripted Fibonacci numbers whose first term is equal to 1 is equal to $\frac{n+1}{2}$.

(ii) Number of consecutive even-subscripted Fibonacci numbers whose first term is equal to 1 is equal to $\frac{n-1}{2}$.

$$S_n = \left(u_{2n+1} - 1\right) + u_{2n}$$

14.11 Validating u_{n+2} - 1 as the Sum of the First n Fibonacci Numbers

If u_{n+2} - 1 is the formula for finding the sum of first n Fibonacci numbers, then

$$S_n = \left(u_{2n+1} - 1\right) + u_{2n} = u_{n+2} - 1$$ has to be a true statement.

Let $\frac{n}{2} = n'$

When n is even $\begin{cases} n' = \frac{n}{2} & \text{for even - subscripted terms} \\ n' = \frac{n}{2} & \text{for odd - subscripted terms} \end{cases}$

When n is odd $\begin{cases} n' = \frac{n-1}{2} & \text{for even - subscripted terms} \\ n' = \frac{n+1}{2} & \text{for odd - subscripted terms} \end{cases}$

14.12 Summing Even Number of Terms of First n Fibonacci Numbers

A. Sum of Even Terms

Sum of even terms is the same as sum of $n/2$ terms

$$S_{(\frac{n}{2})} = u_{2n+1} - 1$$

By substituting for n' we have:

$$S_{(\frac{n}{2})} = u_{2\left(\frac{n}{2}\right)+1} - 1 = u_{n+1} - 1 \quad \text{.......(i)}$$

B. Sum of Odd Terms

Sum of odd terms is the same as sum of $n/2$ terms

$$S_{(\frac{n}{2})} = u_{2n'}$$

By substituting for n' we have:

$$S_{(\frac{n}{2})} = u_{2n'} = u_n \quad \text{.......(ii)}$$

14.13: Summing First *n* Fibonacci Numbers: Another Perspective

$$S_n = S_{(\frac{n}{2})} + S_{(\frac{n}{2})} = S_{(\frac{n}{2}+\frac{n}{2})} = u_{2n'+1} - 1 + u_{2n'} \left(\frac{n}{2}+\frac{n}{2}\right)$$

$$= u_{2(\frac{n}{2})+1} - 1 + u_{2(\frac{n}{2})} = u_{n+1} - 1 + u_n = \left(u_{n+1} + u_n\right) - 1$$

$$= u_{n+2} - 1$$

14.14: Summing Odd Number of First *n* Fibonacci numbers

A. Sum of Even Terms

The sum of even terms is the same as the sum of $\frac{n-1}{2}$ terms.

Therefore, $S_{(\frac{n-1}{2})} = u_{2n'+1} - 1 \quad \left(n' = \frac{n-1}{2}\right)$

By substituting for n' in $u_{2n'+1} - 1$ we have:

$$S_{(\frac{n-1}{2})} = u_{2(\frac{n-1}{2})+1} -1 = u_{n-1+1} -1 = u_n -1 \ldots\ldots\ldots\ldots\ldots\ldots\ldots\ldots(i)$$

B. Sum of Odd Terms

The sum of odd terms is the same as the sum of $\frac{n+1}{2}$ terms.

Therefore, $S_{(\frac{n+1}{2})} = u_{2n'} \quad \left(n' = \frac{n+1}{2}\right)$.

By substituting for n' in $u_{2n'}$ we have:

$$S_{(\frac{n+1}{2})} = u_{2(\frac{n+1}{2})} = u_{n+1} \ldots\ldots\ldots\ldots\ldots\ldots\ldots\ldots\ldots\ldots(ii)$$

We can find the sum of first *n* Fibonacci numbers by adding equation (i) and equation (ii) above. Doing so we have:

$$S_n = (u_n -1 + u_{n+1}) -1 = u_{n+2} -1$$

14.15: Summing First *n* Fibonacci Numbers: Another Perspective

By substituting for n' in $u_{2n'+1}-1$ and $u_{2n'}$ we have:

$$S_n = S_{(\frac{n-1}{2})} + S_{(\frac{n+1}{2})} = S_{(\frac{n-1}{2}+\frac{n+1}{2})}$$

$$= u_{2n'+1} -1 + u_{2n'} \quad \left(\frac{n-1}{2}+\frac{n+1}{2} = n\right)$$

$$= u_{2\left(\frac{n-1}{2}\right)+1} - 1 + u_{2\left(\frac{n+1}{2}\right)}$$

$$= u_{n-1+1} - 1 + u_{n+1} \quad = \quad u_n - 1 \; + \; u_{n+1} \; = \; \left(u_n + u_{n+1}\right) - 1$$

$$= u_{n+2} - 1 \quad \text{.........................(ii)}$$

The result in (i) and (ii) above confirmed the fact that regardless whether n is odd or even, the sum of first n Fibonacci numbers is equal to u_{n+2} - 1.

14.16 Conclusion

We have already simplified and arrived at the following:

$$(a) \; \left(u_{2n+1} - 1\right) + u_{2n} \text{ as } u_{2n+2} - 1$$

$$(b) \; \left(u_{2n+2} - 1\right) \neq u_{n+2} - 1$$

But we found earlier that when n is even, there are:

(a) $n/2$ even-subscripted terms and not n and $n/2$ odd - subscripted terms and not n as Emeka thought. We also found that when n is odd, there are:

(b) $\frac{n-1}{2}$ even-subscripted terms and not n and $\frac{n+1}{2}$ odd-subscripted terms and not n as Emeka also thought. The variable n in $u_{n+2} - 1$ refers to the number of elements in A∪ B (see set C).

Number of elements in set A

$$= \frac{n-1}{2} \text{ or } \frac{n+1}{2}$$

so that $\frac{n-1}{2} - \frac{n+1}{2} = 1$

and $\frac{n-1}{2}+\frac{n+1}{2}=n$

Number of elements in set B

$=\frac{n+1}{2}$ or $\frac{n-1}{2}$

so that $\frac{n+1}{2}-\frac{n-1}{2}=1$ and $\frac{n+1}{2}+\frac{n-1}{2}=n$

What is then our conclusion? The student's logic is wrong. Case closed!

Master Emeka Ibezimako is an imaginary student who exists in the imagination of the author. Any coincidence in real life is unintentional.

14.17 A Search for Counter Examples

Can you think of a subset of first n Fibonacci numbers whose sum is equal to $u_{2n+1}-1+u_{2n}$ where $u_{2n+1}-1$ is the sum of first n consecutive even-subscripted terms and u_{2n} is the sum of first n consecutive odd-subscripted terms?

Can you identify first n odd-subscripted Fibonacci numbers whose sum is not equal to u_{2n}?

Can you identify first n even-subscripted Fibonacci numbers whose sum is not equal to $u_{2n+1}-1$?

Can you identify first n Fibonacci numbers whose sum is not equal to $u_{n+2}-1$?

Appendix

Seminars & Seminar Scheduling

Appendix A

Math-Magic with Paul Chika Emekwulu

A.1 Program Objectives

Develops analytical and logical thinking skills.
Presents the beauty, elegance and excitement in number concepts.
Entertains ad stimulates interest through number tricks and investigatory lessons.
Encourages pattern recognition.
Actively involves and motivates students of all abilities.
Encourages student-teacher, teacher-student, student-student communication.
Supports National Council of Teachers of Mathematics (NCTM) standards.

A.2 Program Description

Math-Magic is neither about magic nor numerology. Math-Magic is a program that uses creative and innovative teaching strategies to make mathematics exciting, interesting, and intriguing to high school students using paper and pencil. Most of these activities are embedded in guided discovery lessons that come in worksheet format.

Math-Magic is about motivation. It is about excitement. It is about mathematical reasoning. It is about pattern recognition. It is not about magic. It is not about numerology.

A.3 Past Engagements

Math-Magic have been presented to the following schools and associations.

Oklahoma Council of Teachers of Mathematics;
Panhandle Mathematics & Science Conference;
Kansas Association of Teachers of Mathematics;
Booker T. Washington High School, Tulsa, OK;
Oklahoma City Community College, Oklahoma City, OK;
Tecumseh High School, Tecumseh, OK;
Jenks Public Schools, Jenks, OK;
Newkirk High School, Newkirk, OK, USA;
Norman High School, Norman, OK, USA;
Oklahoma State University, Stillwater, OK, USA;
Washington High School, Washington, OK, USA;
Oklahoma Education Association;
Liberated Arts Center, Oklahoma City, OK;
Oklahoma State University, (Technical Branch) Oklahoma City, OK;
National Council of Teachers of Mathematics;
Board Members of Organization of Rural Oklahoma Schools

A.4 What People are Saying

"The looks on the students' faces were priceless and the "Ah-Ha's: were abundant when the same equations and theories were introduced in new and interesting ways. We would recommend it to anyone who finds it difficult to grasp math concepts. "Math-Magic" may be the key to unlock the mysteries of the math world." ***Bennie Boykin, Upward Bound Director, Oklahoma State University (Technical Branch), Oklahoma*** **City**

"The Central Regional Conference Meeting of the National Council of Teachers of Mathematics in Topeka, KS will be called a resounding success because of people such as you gave so generously of their time and effort. We are convinced that our conference was among the very best regional conference NCTM has had." ***Dr. Connie S. Schrock, Co-Program Chair, NCTM Central Regional Conference, Topeka, KS***

Dear Paul
National Council of Teachers of Mathematics
Central Regional Conference
St. Louis, Missouri

29-31 January, 1998

"What do 400 excellent speakers, nearly 2400 participants, top notch facilities, well-orchestrated arrangements, spring-like weather and 'show-me" hospitality equal ? A memorable Saint Louis Central Regional Conference of the National Council of Teachers of Mathematics!

The accolades for the quality of the program and local arrangements are still arriving. Your presentation contributed to the praise the Program Committee continues to receive. We appreciate the time, thought, and preparation you gave to your part in the great success of our conference.

Best Wishes and many, many thanks,
Sincerely, Carol A. Edwards, Program Chair

Organization of Rural Oklahoma Schools
Box 189
Foss, OK 73647

March 1, 1995

Paul Chika Emekwulu
Novelty Books
P.O.Box 2482
Norman, OK 73070

Dear Paul,

Thank you for your participation during the January meeting of the OROS board of directors.

The group was very pleased with your presentation. I will be pleased to present members of the OROS board with copies of the "Program Request Form".

It is quite evident that you are able to capture the attention of your audience through "Magic with Numbers".

Sincerely, Tom Butler, Executive Director OROS

The University of Oklahoma
CENTER FOR THE STUDY OF SMALL RURAL SCHOOLS
COLLEGE OF CONTINUING EDUCATION
February 10, 1997

Mr. Paul Chika Emekwulu
Publisher
Novelty Books
P.O.Box 2482
Norman, OK 73070

Dear Mr. Emekwulu

Congratulations! Your presentation entitled session title Math-Magic: Encouraging mathematical reasoning, achieving motivation and excitement in the classroom for the sixth annual National Conference on Creating the Quality School has been reviewed and accepted. The response to the call for presenters has been gratifying. The conference should prove to be exciting and valuable to attendees and presenters alike.

The proposals were reviewed as quickly as possible by a panel to enable presenters ample notification to make travel plans. The conference begins Thursday, March 20 and concludes Saturday, March 22 at 10.30 a.m. (see enclosure).

A full registration brochure is included, remember that registration is required for all presenters. Your pre-registration will help speed up the check-in process at the conference. To register by phone, call 1-800-527-0772 ext. 2248. If you will not be able to attend, please notify us immediately so that we may allow someone else the opportunity to present. If you bring handouts, prepare for approximately 20-25 participants in your session. Please make your session as interactive as possible.

Please be sure your name, the name of your organization, and address at the top of this letter are correct and as you would like them listed in the printed conference program. If there are changes, please write or fax them to us at your earliest convenience.

Please feel free to call us if you have any questions, concerns, or needs. This conference will address important issues, and we look forward to working with you. Again, contact us as needed at 800-937-4760 and ask for Cathie Parker.

See you in March!

Jan C. Simmons
Senior Program Development Specialist
Enclosures

West Texas A & M
UNIVERSITY
Division of Education

August 31, 2000

Paul Emekwulu
P.O.Box 2482
Norman, OK 73070

Dear Paul:

Re: Panhandle Mathematics and Science Conference

Thank you for your proposal for the above upcoming conference on Saturday, September 30th 2000 titled, Math-Magic: Encouraging mathematical reasoning.

I am delighted to inform you that your proposal has been accepted, and that we are in the process of constructing the schedule for the conference at this time. The schedule, with times and room numbers, will be posted to our web site at www.wtamu.edu as soon we have it completed. There is an icon on the bottom left-hand corner of the web page that will lead you to the Panhandle Math/Science Conference site. You will be presenting your session 1 time(s) and we suggest that you prepare for 30 participants in each session.

Registration for the conference will begin at 8 am in the Jack B. Kelley Student Center on the WTAMU campus where there will be a speaker packet waiting for you. Lunch is provided and we hope that you will enjoy your day with us. If for some reason you cannot be with us, please be kind enough to let me know as soon as you can so that arrangements can be made to cancel your session.

If I can be of any further help, please do not hesitate to contact me by email at cpurkiss@mail.wtamu.edu or by phone on 806-651-2618. I look forward to seeing you on the 30th September.

Sincerely, Chris Purkiss
Chairperson

Panhandle Mathematics and Science Conference
A Member of The Texas A & M University System
WTAMU Box 60208 Canyon, Texas 79016-0001 806-651-2626 Fax 806-651-2626

Appendix B

There Could be a Book in You

B.1 Program Objectives

In this seminar, the participants will:

Gain inspiration and motivation that could translate into action.
Realize that dreams and intuition could be sources of book ideas.
Realize that only two things can stop a dream book.
Realize that affirmations could be used to activate our creativity.

B.2 Program Description

How many of you have ever thought of writing a book?

Are you working on a book now or have you ever submitted a manuscript to a publisher? Do you think you have a special knowledge, or skills you would like to share with others? Are you a good story-teller? Can you tell stories in a manner which holds people's attention?

Are you a teacher and you have unusual but creative ways of presenting ordinary classroom concepts?

Are you currently doing seminars based on your experiences, and you don't have a book covering your topics?

Do you write articles for newspapers or magazines?

If yes, have you ever thought of building a book out of these articles?

Do you have an idea for a book but don't know how to put them together?

Have you ever thought of collaboration with someone on a book project?

Do you have any ideas or particular cause you would like to be remembered for?

If your answer to any of the above questions is yes, you need to write a book.

Emekwulu maintains that only two things can stop anyone from writing his or her dream book. Lack of faith in the message and lack of faith in the messenger.

B.3 Past Engagements

Norman Galaxy of Writers Inc.;
Oklahoma City Writers Inc.;
Canadian Valley Lions Club;
Metropolitan Library System, Mid-West City, OK;
Elk City Carnegie Friends of the Library, Elk City, OK;
Mid-Oklahoma Writers' Club;
Oklahoma Education Association;
Moore Association of Classroom Teachers, Moore; OK.

B.4 What People are Saying

Thank you, Paul, for your encouragement—Norman Galaxy of Writers

Appendix C

Miscellaneous Exercises

1. For what values of n is $n^2 - n = n^2 + n$?

    A. 2
    B. 1
    C. -1
    D. None of the above

2. How many odd numbers are there between 1 and 40?

    A. 99
    B. 97
    C. 38
    D. None of the above

3. If the sum of first n odd integers is 256, find the last term.

    A. 16
    B. 31
    C. 257
    D. None of the above

4. The rule for finding the sum of first n even numbers is given by:

   A. $n^2 - n + 1$
   B. $n^2 - n$
   C. n^2
   D. None of the above

5. When expressed in expanded form, *abc* becomes:

   A. $10 + b + c$
   B. $100a + b + c$
   C. $100a + 10b + c$
   D. None of the above

6. If n is an odd number, then

   A. $n(n - 1)$ is even
   B. $n + 2$ is even
   C. n is even
   D. All of the above

7. If a is an even integer, the following statements are true except:

   A. a^2 is even
   B. $2a^2$ is even
   C. $a - 2$ is even
   D. $a^2 - 1$ is even

8. The rule for finding the sum of first n positive even numbers whose first term is 2 is given by:

   A. $n^2 + 2n$
   B. $n^2 + n$
   C. $n^2 - n$
   D. None of the above

9. If $5a = 5b + c$ and $a = b$, the numerical value of c is:

   A. 0
   B. $5a - 5b$
   C. 5
   D. None of the above

10. The numbers a and b ($b > a$) are two consecutive positive odd numbers. If $a + b = 24$, and $ab = 143$, what is $b^2 - a^2$?

   A. 2(13-1)
   B. 2(13+34)
   C. 2(13+11)
   D. None of the above

11. Study the following table and answer the question that follows. Which of the following describes the relationship between h and w?

h	***w***	***h - w***
9	3	6
7	5	2
6	1	5
5	2	3
4	2	2

   A. $h - w = 2$
   B. $h - w < 2$
   C. $h - w \geq 2$
   D. None of the above

12. The following numbers are all divisible by 9 except:

   A. 81239124564
   B. 99269862650
   C. 20248126767
   D. None of the above

13. The last term of a set of first n positive odd integers where a is the first term and n is number of terms is given by:

A. $a + 2(n - 1)$
B. $a + (n - 1)$
C. $a + (n + 1)$
D. None of the above

14. If the sum of a set of numbers is given by $n^2 + n$, then that set is a set of:

A. first n even numbers
B. first n even numbers whose first term is 2
C. first n even numbers, the first term notwithstanding
D. None of the above

15. For any four consecutive odd numbers a, b, c, d, the following are true except:

A. $a - c = -4$ and $c - b = 2$
B. $c - a = 4$ or $d - c = 4$
C. d - b = c - a
D. None of the above

16. What number must be added to each term of the ratio, 3: 8 so that the ratio becomes 3:5?

A. 4.5
B. 11
C. 8
D. None of the above

17. $a/a = 1$ if and only if the following are true:

A. a = 1
B. a = 0
C. a ≠ 0
D. None of the above

18. The general form of a quadratic equation is given by $ax^2 + bx + c = 0$, $a \neq 0$, $y = ax^2 + bx + c$ is an equation of a straight line if:

    A. $a = 0, b = 0$ or $a = 0, b \neq 0, c = 0$ or $a = 0, b \neq 0$ and $c \neq 0$
    B. $a = 0, b = 0, c = 0$
    C. $a \neq 0, b = 0, c = 0$
    D. None of the above

19. The difference between a Fibonacci number x and a triangular number y is 167 and their sum is 299. Find $2x + y$.

    A. 532
    B. 344
    C. 598
    D. 132

20. The sum of a set first n whole numbers is 28. How many whole numbers are in this set?

    A. 14
    B. 7
    C. 3
    D. None of the above

21. The following numbers belong to the same set of special whole numbers except:

    A. 100
    B. 25
    C. 36
    D. 48

22. The ratios of two successive Fibonacci numbers approach a constant number. That number is:

A. 1.6182
B. 1.6180
C. 2.7183
D. None of the above

23. The last term of a set of first n positive odd integers is given by $a + 2(n-1)$. 2 is known as:

A. Common difference
B. Odd difference
C. Common ratio
D. None of the above

24. For three consecutive odd numbers, a, b and c, $c^2 - a^2 = 184$, find the largest of the three odd numbers.

A. 19
B. 21
C. 25
D. None of the above

25. If $\frac{8}{9}xy = \frac{8}{9}$, find the value of xy.

A. $\frac{8}{9}$
B. y
C. 1
D. None of the above

26. The last term of a set of first n positive integers is given by $a+2(n-1)$. If this set is a set of first n positive odd integers, then d is equal to:

A. 1
B. 2
C. 4
D. None of the above

27. For any four consecutive whole numbers *a, b, c, d,*

$$\frac{(d+a)(d-a)}{b+c}=3$$

Find *d* in terms of *a, b, c.*

A. $\sqrt{a^2+6b+6c}$
B. a + 3b + 3c
C. 3(b+c)
D. $\sqrt{a^2+3b+3c}$

28. The last term *L* of a set of first n positive numbers is given by $L=a+(n-1)d$. If this is the set of first n whole numbers, then *d* is:

A. -1
B. 1
C. 2
D. None of the above

29. For any four consecutive odd numbers *a, b, c, d,* if $a+b+c+d=2(b+c)$, then the following are true except:

A. $a+d=b+c$
B. $c-a=d-b$
C. $a=c-d+b$
D. None of the above

30. Given that a, b, c, d, e are five consecutive Fibonacci numbers, e-a = 2b+c and $b + d = e\text{-}a$, then the following is always true:

    A. $b+d = b+c$
    B. $b+d = 2b+c$
    C. $b+d = 2c+1$
    D. None of the above

31. If a and b are two consecutive whole numbers, then $a + b$ in terms of a is equal to:

    A. $2a+1$
    B. $2a - 1$
    C. $2a$
    D. None of the above

32. For any four consecutive odd numbers a, b, c, d, if $2(c - 1) = a + d$, and $b + c = a + d$, then:

    A. $2(c - 1) = b + c$
    B. $2(c - 1) = 2(a + d)$
    C. $2(c - 1) = 2(b - c)$
    D. None of the above

33. If a and b are two consecutive whole numbers, then a + b in terms of b is equal to:

    A. 2b
    B. 2b-1
    C. 2b + 1
    D. None of the above

34. If a, b, c, d are four consecutive odd numbers, $3c = b + c + d$ and $a + d = b + c$, then a+ d is equal to:

A. $3c - d$
B. $3(c - d)$
C. $3c + d$
D. None of the above

35. If a, b, c are three consecutive odd numbers and $ac = 77$, what is b?

A. 9
B. 18
C. 81
D. None of the above

36. Given that $a+b+c = \frac{1}{2}(a+L)$ for any three consecutive odd numbers a, b, c, find b in terms of a and c.

A. $2(a + c)$
B. $\frac{a+c}{2}$
C. $a + c$
D. None of the above

37. The number of diagonals in a nine-sided polygon is:

A. 27
B. 54
C. 45
D. None of the above

38. If $nq = 4q$, what is n?

A. 2 or -3
B. 2
C. 8
D. None of the above

39. If $(d + b)(d - b) = 8c$ for any three consecutive odd numbers, then:

A. $d - b = 3$
B. $d + b = 2c$
C. $d + b = 4c$
D. None of the above

40. a, b, c are three consecutive odd integers. If $\frac{b}{2} = \frac{c+a}{c-a}$, what is $\frac{1}{b^2}$?

A. $\left|\frac{1}{2}(c+a)\right|^2$

B. $\frac{c+a}{2}$

C. $\frac{4}{c^2 + 2ac + a^2}$

D. None of the above

41. If $c^2 - a^2 = 8b$ for any three consecutive odd numbers, then:

A. $c - a = 2c$
B. $c - a = c + a$
C. $c + a = 2b$
D. None of the above

42. If a, b, c, d are four consecutive odd numbers, the following are not true except:

A. $a + d = 2c$
B. $a + 2d = 3c$
C. $a - d = 2c$
D. None of the above

43. If L and SUM are the last term and sum of first n positive odd integers respectively, and $L = 2\sqrt{SUM} - 1$, the following is true:

A. $L^2 - 2L + 1 = 0$
B. $L - 2n + 1 = 0$
C. $n = L^2 - 2L + 1 = 0$
D. None of the above

44. If L and SUM are the last term and sum of first n positive odd integers respectively, and $L = 2\sqrt{SUM} - 1$, the following is true:

A. $n = L + 1$
B. $n = L^2 + 1$
C. $n = L - 1$
D. $L^2 = 4n^2 - 4n + 1$

45. If $b^2 - a^2 = 4(a + 1)$ for any two consecutive odd numbers, then an expression for $4a + 1$ is:

A. $4b - 7$
B. $4b + 4$
C. $4b - 2$
D. None of the above

46. If $n = \frac{L+1}{2}$ for a set of first n positive odd integers (L = last term), then $L - 1$ is always:

A. even
B. odd
C. prime
D. None of the above

47. If a polygon has 54 diagonals, then it is described as a:

A. Nonagon
B. Duocagon
C. Decagon
D. None of the above

48. A set of consecutive odd numbers has the first term as 49 and its last term as 109. How many odd numbers are in the set?

A. 118
B. 100
C. 31
D. None of the above

49. If $2(c + d) = a + b + c + d$ for any four consecutive odd numbers a, b, c, d, then:

A. $a + d = b + c$
B. $2a + d = b + c$
C. $2a - d = b$
D. None of the above

50. The sum of a set of first n positive even numbers whose first term is 0 is given by $n^2 - n$. If such a set has a sum of 10,100, then the last term is:

A. 200
B. 10,100
C. 10,000
D. None of the above

51. For a set of consecutive even numbers whose first term is 2, the last term is given by:

A. $\frac{1}{2}n$
B. $2n$
C. $4n$
D. None of the above

52. If the set of first n positive odd integers add up to 1024, then the number:

A. 8
B. 16
C. 32
D. None of the above

53. If the set of first n positive odd integers add up to 1024, then the last term of this set is:

A. 512
B. 63
C. 1024
D. None of the above

54. One of the following represents three consecutive odd numbers where a is an odd number greater than or equal to 7.

A. $a - 6, a - 4, a - 2$
B. $a - 6, a - 2, a - 4$
C. $a - 4, a + 2, a$
D. None of the above

55. Which one of the following represents four consecutive whole numbers where a is equal to 0?

A. $a, a + 1, a + 2, a + 3$
B. $a, a + 3, a + 2, a + 1$
C. $a, a - 3, a - 2, a + 1$
D. None of the above

56. If $n \geq 1$ then the following is odd except:

A. $2n - 1$
B. $2n - 2$
C. $2n+1$
D. $4n+1$

57. If a and b are consecutive odd numbers, then $b^2 - a^2$ is equal to:

A. $2a - 1$
B. $2(a+b)$
C. $a + b$
D. None of the above

58. The fractions $\frac{3}{4}$ and $\frac{6}{8}$ are said to be

A. Unit fractions
B. Equivalent fractions
C. Improper fractions
D. None of the above

59. If $x/12 = 3/9$, what is the reciprocal of x?
   A. 24
   B. 36
   C. 0.25
   D. None of the above

60. If $3\frac{1}{2}y = 3\frac{1}{2}$, then y:

   A. 14
   B. 7
   C. 5
   D. None of the above

61. All of the following are false except:

   A. The lcm of any three consecutive positive even integers is equal to their product.
   B. The lcm of any three consecutive positive odd integers is equal to their product.
   C. The sum of three Fibonacci numbers is always even
   D. None of the above

62. All these are Fibonacci numbers except:

   A. 233
   B. 377
   C. 12
   D. None of the above

63. If u_n represents a Fibonacci number, then the next Fibonacci number is represented by:

A. $u_n - 1$
B. u_{n+1}
C. u_{n+2}
D. None of the above

64. For any three consecutive Fibonacci numbers *a, b, c,*

A. $b^2 - ac = 1$
B. $b^2 - ac = \pm 1$
C. $b^2 + ac = 0$
D. None of the above

65. If u_n represents a Fibonacci number, then the following is a set of three consecutive Fibonacci numbers:

A. u_n, u_{n-1}, u_{n+2}
B. $u_{n+1}, u_{n-1}, u_{n+2}$
C. u_n, u_{n+2}, u_{n-1}
D. None of the above

66. If u_n represents a Fibonacci number, then the following is true:

A. $u_{n+2} = u_{n-1} + u_{n+1}$
B. $u_{n+2} = u_n + u_{n-1}$
C. $u_{n+2} = u_{n+1} + u_n$
D. None of the above

67. For any three consecutive Fibonacci numbers *a, b, c,*

A. $a^2 = b^2 + c^2$
B. $c^2 = a^2 + b^2$
C. $c = a + b$
D. None of the above

68. The smallest number that can be added to 24,671 to make it divisible by 3 is:

A. 6
B. 3
C. 1
D. None of the above

69. The product of the LCM and HCF of two numbers is 24. If one of the numbers is 6, the other number is:

A. 10
B. 12
C. 2
D. None of the above

70. If half the product of the LCM and HCF of two numbers is 144, and one of the numbers is 12, the other number is:

A. 144
B. 24
C. 18
D. None of the above

71. For any two whole numbers a and b, where a > 0 and b > 0, the following is true:

    A. (LCM of a and b) × (HCF of a and b) = ab
    B. (LCM of a and b) × (HCF of a and b) = a + b
    C. (LCM of a and b) × (HCF of a and b) = b-a
    D. None of the above

72. *a, b, c* are three consecutive odd numbers. If twice the middle plus the largest is equal to 131, then the smallest is:

    A. 41
    B. 21
    C. 3
    D. None of the above

73. The sum of two odd numbers is 32 and their difference is 14. The product of the numbers is:

    A. 18
    B. 46
    C. 207
    D. None of the above

74. The product of two odd numbers is 207. If their difference is 14, then half the smallest of these numbers is:

    A. 9
    B. 4.5
    C. 23
    D. None of the above

75. The least number divisible by both 6 and 8 with a remainder of 3 is:

A. 45
B. 48
C. 27
D. None of the above

76. If two numbers a and b ($b > a$) are consecutive odd numbers, then the following is true:

A. $b + a = 2$
B. $b - a = a - b$
C. $b - a = -2$
D. None of the above

77. If the HCF of two whole numbers a and b is 1, then a and b are said to be:

A. Evenly prime
B. Relatively prime
C. Consecutive
D. None of the above

78. If $a < 0$ and $b < 0$, then:

A. $\frac{a}{b} = 0$

B. $\frac{a}{b} < 0$

C. $\frac{a}{b} > 0$

D. None of the above

79. Two consecutive triangular numbers can be represented generally as one of the following:

A. $\frac{n(n+1)}{2}$ and $\frac{n^2 + 3n + 2}{2}$
B. $\frac{n(n+1)}{2}$ and $\frac{n^2 + 3n}{2}$
C. $\frac{n(n+1)}{2}$ *or* $\frac{n^2 + 3n + 2}{2}$
D. None of the above

80. The 100th triangular number is:

A. 4050
B. 2525
C. 5050
D. None of the above

81. The last term of a set of first n odd numbers is given by $2n+1$. This is true if and only the first term is:

A. 3
B. 2
C. 1
D. None of the above

82. The sum of a set of first n positive odd integers is given by n^2. This is true if and only if the first term is:

A. 2
B. 0
C. 3
D. None of the above

83. If *bc* is divisible by 2, then:

A. 10b and c are both divisible by 2
B. Only c is divisible by 2
C. $b + c = 10$
D. None of the above

84. If *abcd* is divisible by 5, then one of the following is true:

A. Either $d = 0$ or $d = 5$
B. d cannot be determined from available information
C. $a + b + c + d = 5$
D. None of the above

85. If abcd is divisible by 4, then the following is true:

A. $ac + d$ is a multiple of 4
B. $a + b + c + d < 4$
C. $a + b + c + d = 4$
D. None of the above

86. If *abc* is divisible by 4, then

A. $a = 4$
B. bc is always a multiple of 4
C. b is always less than 4
D. None of the above

87. $abcd - (a + b + c + d)$ is equal to:

A. $9(111a+11b + c)$
B. $999a + 9ab$
C. Meaningless
D. None of the above

88. If xyz is divisible by 2, then the smallest value x can take is:

A. 0
B. 1
C. 2
D. None of the above

89. If pqr is divisible by 2, the maximum value r can take is:

A. 0
B. 10
C. 8
D. None of the above

90. If $pqrstuv$ is divisible by 11, then one of the following is true:

A. $(p + r + v) - (q + s + u) = 0$
B. $(p + r + v) - (q + s + u) = 0$
C. $(p + r + v) - (q + s + u) = 0$
D. None of the above

91. For a set of n consecutive even numbers whose first term is 2, the last term is given by one of the following:

A. $\frac{n}{2}(n^2 + 2n)$

B. $2n^2 + 2n$
C. $n^2 - n$
D. None of the above

92. The sum of n consecutive even number even numbers whose first term is 2 is given by $n^2 + n$. If such a set has a sum of 101,00, the last term is:

A. 200
B. 10,100
C. 10,000
D. None of the above

93. If $n(n-1)$ is even and $n > 0$, then the following are true except:

A. If n is even, $n - 1$ is even
B. If n is odd, $n - 1$ is even
C. $n(n - 1) - n^2$ is negative
D. None of the above

94. All these are rational numbers except:

A. $\sqrt{3}$

B. 4

C. $\frac{1}{2}$

D. None of the above

95. If $\frac{4}{4-\sqrt{7}} = a - b\sqrt{c}$, a is equal to:

A. -16

B. $\frac{-16}{5}$

C. $\frac{1}{\sqrt{7}}$

D. None of the above

96. $\frac{4}{\sqrt{7}-\sqrt{3}} = \frac{a}{\sqrt{7}+\sqrt{3}}$, then a is equal to:

A. $10 + 2\sqrt{21}$

B. $\sqrt{10} + \sqrt{21}$

C. $\sqrt{7} - \sqrt{3}$

D. None of the above

97. In the expression, $\frac{\sqrt{5}+\sqrt{3}}{8+2\sqrt{15}}$, the denominator is:

A. Rational
B. Irrational
C. An integer
D. None of the above

98. What must be added to x^2 - $6x$ to make it a perfect square?

A. 4.5
B. 18
C. 3
D. None of the above

99. Arranged in ascending order, the set of fractions, $\frac{1}{2}, \frac{1}{5}, \frac{1}{3}, \frac{1}{4}$ becomes:

A. $\frac{1}{2}, \frac{1}{3}, \frac{1}{4}, \frac{1}{5}$

B. $\frac{1}{3}, \frac{1}{2}, \frac{1}{4}, \frac{1}{5}$

C. $\frac{1}{5}, \frac{1}{4}, \frac{1}{3}, \frac{1}{2}$

D. $\frac{1}{4}, \frac{1}{5}, \frac{1}{2}, \frac{1}{3}$

100. For any polyhedron, the following is true:

A. $F + V - E = 2$
B. $F + E - V = 3$
C. $F + V - E = 3$
D. None of the above

where F = number of faces, V = number of vertices, E = number of edges

101. If *a, b, c, d* are consecutive odd numbers, the following are true except:

A. $a + b > c + d$
B. $a + b > 2c - d$
C. $a + b = d - b$
D. $b + c = a + d$

102. For any set of first *n* whole numbers, the last term *L* is given by one of the following:

A. $a+(n-1)d$
B. $a + 2$
C. $a + 3$
D. None of the above

where a = first term, n = number of terms, and d = common difference.

103. 1, 3, 5, 7, 9, . . . is a set of first *n* positive odd integers. The 100th term of the set is:

A. 200
B. 101
C. 197
D. 199

104. What is the best prediction for the 20th term of the sequence:
$1^2 + 3^2 + 5^2 + 7^2$ .....

A. 39^{40}
B. 38^2
C. 39^{38}
D. None of the above

105. What is the best prediction for the 100th term of the sequence:
$1^2 + 3^2 + 5^2 + 7^2$ .....

A. 199^2
B. 198^2
C. 100^2
D. None of the above

106. If $b^2 - a^2, = 4(a+1)$, for any two consecutive odd numbers, then the following statements are not true except:

A. $\frac{1}{2}(b+a) = a+1$

B. $\frac{1}{2}b = \; y+1$

C. $b = y + 1$

D. $\frac{1}{2}(b+a) = y + 2\backslash$

107. If $\frac{1}{2}(b+a) = 7$ and $y + 2 = 4 + 3$, then these statements are false except:

A. b = y+1

B. $\frac{1}{2}b = y+1$

C. b = y + 1

D. $\frac{1}{2}(b+a) = y+2$

108. If b^2-xy = 15xy, where x and y are real numbers, solve for b. The equation has the following number of real solutions:

A. 1
B. 3
C. 2
D. 0

109. Find the 12th term in the sequence 10, 14, 18 . . .

A. 27
B. 54
C. 34
D. 10

110. Find the 60th term of an arithmetic sequence if a = 7.5 and d = 0.5.

A. 22
B. 37
C. 59
D. 29

Appendix C

Solutions to Miscellaneous Exercises

1. D	23. A	45. A	67. C	89. C
2. D	24. C	46. A	68. C	90. D
3. B	25. C	47. B	69. D	91. C
4. B	26. B	48. C	70. B	92. A
5. C	27. D	49. A	71. A	93. A
6. A	28. B	50. A	72. A	94. A
7. D	29. D	51. B	73. C	95. D
8. B	30. B	52. C	74. B	96. A
9. A	31. A	53. B	75. C	97. B
10.C	32. A	54. A	76. D	98. D
11.C	33. B	55. A	77. B	99. C
12.D	34. A	56. B	78. C	100.A
13.A	35. A	57. B	79. A	101.B
14.B	36. B	58. B	80. C	102.A
15.B	37. A	59. C	81. A	103.D
16.A	38. D	60. D	82. D	104.D
17.C	39. B	61. B	83. A	105.B
18.A	40. C	62. C	84. A	106.A
19.A	41. C	63. B	85. D	107.D
20.D	42. B	64. B	86. B	108.C
21.D	43. B	65. D	87. A	109.B
22.B	44. D	66. C	88. A	110.C

Appendix D

Trial Questions on Numbers of the Fibonacci Sequence

1. Prove that for any five consecutive Fibonacci numbers a, b, c, d, e: $e - a = 2b+c$.

2. Given that $b^2 - ac + 1 = 0$ and $a^2 - 3ac + c^2 + 1 = 0$, find an expression for $b + c$.

3. Prove that for any four consecutive Fibonacci numbers *a, b, c, d*: $\sqrt{a^2 + 4bc}$ is always a perfect square whose square root is equal to d.

4. Given that $a^2 + c^2 - 3ac = 1$, find c in terms of a.

5. Prove that $c^2 - 2ac + a^2 = ac - 1$ when a and c have odd subscripts.

6. Given that $e - a = b + d$, prove that $e - a = 2b + c$.

7. For any five consecutive Fibonacci numbers *a, b, c, d* and *e*, prove that $be - cd = \pm 1$.

8. If $a^2 + 4bc = 64$, and *a, b, c* are consecutive Fibonacci numbers, solve for *a, b,* and *c*.

9. Answer true or false to the following questions:

   (a) The sum of any two Fibonacci numbers is always a Fibonacci number.
   (b) The sum of any two Fibonacci numbers is not necessarily a Fibonacci number.
   (c) For values of a and c whose subscript are even, $b^2 - ac = \pm 1$ where a, b, c are three consecutive Fibonacci numbers.
   For any three consecutive Fibonacci numbers a, b, c, $b^2 + 2ab$ is always a Fibonacci number where a and c have even subscripts.
   (d) The product of two or more Fibonacci numbers can never result to a Fibonacci number.

10. For any four consecutive Fibonacci numbers a, b, c, d, prove that $c^2 - bd = \pm 1$.

11. If $a^2 + 4bc = 64$, find a in terms of b and c.

12. If $a^2 + 4bc = (b+c)^2$, what is $a^2 + 2bc$?

13. Find the limit of $\dfrac{Z_n}{Z_{n+1}}$ as n approaches infinity where

    $Z_n = (1,3,7,13,21,31,43,57,73)$.

14. If $b^2 - ac + 1 = 0$, then $a^2 - 3ac + c^2 + 1 = 0$. Find a similar expression for $b^2 - ac - 1 = 0$ where a and c have odd-subscripts.

15. For any three consecutive odd or even numbers a, b, c, $3b = a + b + c$. Is this true for any three consecutive Fibonacci numbers?

16. Given that u_n is a Fibonacci number, prove that $u_n^2 + u_{n+1}^2$ is a Fibonacci number with an odd subscript.

17. If the equation, $ax^2 - bx - c = 0$ if and only if a, b, c are consecutive Fibonacci numbers, prove that the sum of roots and product of roots of the above equation are b/a and $-c/a$ respectively.

18. The sum of twice a triangular number and its subscript is 24. If their difference is 6, find the triangular number.

D.1 Solutions To Odd-Numbered Questions

1. e-a = (c + d) - a
= (a +b) + (b + c) - a
= (a-a) + (b + b) + c = 2b + c

3. $a^2 + 4bc = (c-b)^2 + 4bc$
$= c^2 - 2bc + b^2 + 4bc$
$= c^2 + 2c + b^2 = (c+b)^2 = (b+c)^2 = d^2$

Taking the square root of both sides we have:

$\sqrt{(c+b)^2} = \sqrt{d^2}$

$c + b = d$
Since d is a perfect square, and

$\sqrt{a^2 - 4bc} = d^2$, and $\sqrt{d^2} = d, \sqrt{a^2 + 4bc}$ is a perfect square whose square root is equal to d.

5. Since c = a + b, $c^2 - 2ac + a^2 = (a+b)^2 - [2a(a+b)] + a^2$

$(a+b)^2 - [2a(a+b)] + a^2 = a^2 + 2ab + b^2 - (2a^2 + 2ab) + a^2$

$= (a^2 + a^2) - 2a^2 + 2ab - 2ab + b^2 = b^2$

But $b = \sqrt{ac+1}$.

Squaring both sides we have:

$b^2 = ac + 1$

But $c^2 - 2ac + a^2 = b^2$

Therefore, $c^2 - 2ac + a^2 = ac + 1$ if and only if a and c have even subscripts since ac+1 is equal to b^2 if a and c have even subscripts.

7. $be - cd = b(c+d) - cd = bc + bd - cd$

$bc + bd - cd = bc + bd - d(a+b)$

$= b(a+b) + b(b+c) - ad - bd$

$= ab + b^2 + b^2 + bc - ad - bd$

$= ab + b^2 + b^2 + bc - a(b+c) - b(b+c)$

$= ab + b^2 + b^2 + bc - ab - ac - b^2 - bc$

$= (ab - ab) + (b^2 - b^2) + b^2 + (bc - bc) - ac = b^2 - ac$

For any three consecutive Fibonacci numbers a, b, c, b^2 - ac =±1.

Therefore, be - cd = ±1, since be - cd = b^2 - ac.

9. **(a)** False **(b)** True **(c)** False **(d)** True **(e)** False

11. $a^2 + 4bc = 64\infty$

$a^2 = 64 - 4bc$

$a = \sqrt{64 - 4bc}$

$= \sqrt{4(16 - bc)}$

$= 2\sqrt{16 - bc}$

Therefore, a = $2\sqrt{16 - bc}$

13. $Z_n = n^2 + n + 1$

$Z_{n+1} = (n+1)^2 + n + 1 + 1$

$= n^2 + 2n + 1 + (n+1) + 1$

$= n^2 + 3n + 3$

Therefore, $\lim_{x\to\infty} \frac{Z_n}{Z_{n+1}} = \lim_{x\to\infty} \frac{n^2 + 3n + 2}{n^2 + n + 1}$

$$= \lim_{x\to\infty} \frac{\frac{n^2}{n^2} + \frac{1}{n^2} + \frac{n}{n^2}}{\frac{n^2}{n^2} + \frac{n}{n^2} + \frac{1}{n^2}} = \lim_{x\to\infty} \frac{1 + \frac{3}{n} + \frac{3}{n^2}}{1 + \frac{1}{n} + \frac{1}{n^2}}$$

As $n \to \infty$, $\frac{3}{n} \to 0, \quad \frac{3}{n^2} \to 0, \quad \frac{1}{n} \to 0$

Therefore, $\lim_{n\to\infty} \frac{1 + \frac{3}{n} + \frac{3}{n^2}}{1 + \frac{1}{n} + \frac{1}{n^2}} = \frac{1+0+0}{1+0+0} = 1$

Therefore, $\lim_{n\to\infty} \frac{Z_{n+1}}{Z_n} = 1$

15. No

17. $ax^2 + bx + c = 0$

Dividing throughout by a we have:

$$x^2 - \frac{b}{a}x - \frac{c}{a} = 0$$

Transposing terms in the above equation we have:

$$x^2 - \frac{b}{a}x = \frac{c}{a}$$

Adding square of half the coefficient of x we have:

$$x^2 - \frac{b}{a}x + \left(\frac{b}{2a}\right)^2 = \frac{c}{a} + \left(\frac{b}{2a}\right)^2$$

Re-writing the above we have:

$$\left(x + \frac{b}{2a}\right)^2 = \frac{b^2}{4a^2} + \frac{c}{a} = \frac{b^2 + 4ac}{4a^2}$$

Taking the square root of both sides we have:

$$x + \frac{b}{2a} = \sqrt{\frac{b^2 + 4ac}{4a^2}}$$

$$= x + \frac{b}{2a} = \frac{\sqrt{b^2 + 4ac}}{2a}$$

Subtracting $\frac{b}{2a}$ from both sides of equation we have:

$$\mathrm{x} + \frac{\mathrm{b}}{2\mathrm{a}} - \frac{\mathrm{b}}{2\mathrm{a}} = \frac{-\mathrm{b}}{2\mathrm{a}} + \frac{\sqrt{\mathrm{b}^2 + 4\mathrm{ac}}}{2\mathrm{a}}$$

$$x = \frac{-b \pm \sqrt{b^2 + 4ac}}{2a}$$

Let the two roots of the equation be x_1 and x_2.

Therefore, if $x_1 = \frac{-b - \sqrt{b^2 + 4ac}}{2a}$ then

$$x_2 = \frac{-b + \sqrt{b^2 + 4ac}}{2a}$$

Sum of roots = $x_1 + x_2$

$$= \left\{\frac{(-b - \sqrt{b^2} + 4ac)}{2a} + \frac{(-b + \sqrt{b^2} + 4ac)}{2a}\right\}$$

$$-\frac{(-b-\sqrt{b^2+4ac})+(-b+\sqrt{b^2+4ac})}{2a}$$

$$\frac{-(b+b)-\sqrt{b^2+4ac}+\sqrt{b^2+4ac}}{2a}$$

$$=\frac{-2b}{2a} = \frac{-b}{a}$$

Product of roots = $(x_1)(x_2)=\left[\frac{-b-\sqrt{b+4ac}}{2a}\right]\left[\frac{-b+\sqrt{b+4ac}}{2a}\right]$

$$=\frac{(-b)^2-\left(b^2+4ac\right)}{4a^2}$$

$$=\frac{4ac}{4a^2} = \frac{c}{a}$$

Appendix E

Paul Emekwulu's Internet Presence

Befriend us on Facebook
http://www.facebook.com/people/Paul-Emekwulu/100000584321587

Read my blogs
http://www.emekwulu.blogspot.com/

Visit my website
http://www.paulemekwulu.com

Follow me on Twitter
http://www.twitter.com/pemekwulu

Read my articles
http://triond.com/users/pemekwulu

Read my articles
http://hubpages.com/author/pemekwulu/best/

E-Mail: patpaul@paulemekwulu.com

Index

I

J

K

L

M

Q

R

S

W

www.ingramcontent.com/pod-product-compliance
Lightning Source LLC
Chambersburg PA
CBHW031818170526
45157CB00001B/108

* 9 7 8 1 4 5 3 5 5 1 0 2 8 *